AF873729

The Last Challenge Problem: George Boole's Theory of Probability

by

David W. Miller

Copyright David W. Miller 2009-2012.

All rights reserved. No part of this publication may be reproduced, distributed, or transmitted in any form or by any means, including photocopying, recording, or other electronic or mechanical methods, without the prior written permission of the publisher, except in the case of brief quotations embodied in critical reviews and certain other noncommercial uses permitted by copyright law.

This book is dedicated

to

REBECCA NICOLE MILLER

August 27, 1962 - January 2, 1968

who was cheated of everything

and to whom I could give nothing else

PREFACE

A preface is the place where an author traditionally offers excuses and apologies, extends thanks, professes gratitude, and often suggests some justification for the book that follows. I have no reason to diverge from that tradition except that I will reverse the order.

When I started to work out Boole's theory of probability quite a few years ago I was familiar with the fact that this theory was considered to be wrong by everyone who commented on it. The theory as Boole presented it is remarkably difficult and, frankly, today as I look back on my efforts I wonder why I persisted. Mostly, I was sustained by the belief that a person of Boole's level of genius simply could not have created a theory quite as bad as the comments about it would suggest. It therefore gives me considerable pleasure to demonstrate in this book that Boole's theory is not only consistent and reasonable - it actually works on real-life problems. In this same context I am pleased to be able to show that Boole's method is equivalent, in a large class of real-world problems, to the application of Jaynes' Principle, which has, seemingly, a completely different logical underpinning.

When I began writing this book I was unaware of any book dealing with Boole's theory of probability since John Maynard Keynes's "A Treatise on Probability", published in 1921. I had not discovered the first edition (1976) of Theodore Hailperin's fine book, "Boole's Logic and Probability".

I became aware of the second edition (1986) when I was well into writing a preliminary version of the present book. My initial trepidation that my own work might already have been done disappeared when I found that Professor Hailperin's goals were quite different than mine. I have devoted a Section of the present book to my comments about his book so I need not discuss it further here.

I was pleased to discover Hailperin's book because it showed me that there were people still interested in the remarkable structure of Boole's thought. When one concentrates attention on the title of Boole's opus major, "An Investigation of the Laws of Thought on Which are Founded the Mathematical Theories of Logic and Probabilities", it is evident that he considered his theory of probability to be on an equal level with his theory of logic. Many early commentators thought that Boole developed his logical theory precisely so he could use it in his probability theory. This granted, there has been more than 100 years in which there has been no positive assessment or use of his probability theory whatsoever. I certainly do not know of any comparable instance where the major contribution of a great thinker has been so consistently ignored.

Of course, there have always been a few lovers of Boole - although it seems to be very few. One such person that I must mention is Herbert F. Ayres. His untimely death was a great loss to the management science profession. He had one of the finest minds I have ever known and he was devoted to Boolean thinking in every area. The very many hours of discussion I had with Herb about the Boolean approach to just about everything were a major reason why I was sure there had to be a meaningful probability theory in Boole's structure. I feel a very deep sense of loss in not having the opportunity to have Herb comment on this book.

I must express my thanks to my family. My daughter, Rachael, helped me with design questions, including the title page. My son, Jason, handled all of the computer and printer hook-ups that permitted me to finish this book and he also taught me how to use them. Finally, my wife, Lisa Evered, offered constant encouragement and help as well as providing a sympathetic ear when my conceptualizations crumbled.

Finally, I should make a few comments about the publication of this book. Mentioned in the third paragraph is that I was working on a preliminary version of this book in the late 1980's. I was quite dissatisfied with my attempt - for example, I had not yet worked out the relationship between Boole's method and Jaynes' Principle - and the ordinary banalities of life have prevented me from producing a more satisfactory version until now. This, therefore, is the first edition.

Probably I should - and I certainly will - offer an apology for the fact that I am publishing this book myself. This automatically means that there will be no professional distribution and, hence, some persons who might have wanted to read this book will not have the opportunity. I will mention that I have had quite a few books published over the years but in the present instance I considered that writing this book was a labor of love, overused though that phrase is. I wanted the price to be as low as possible so that no one would be prevented from reading the book because of its cost. Therefore, I decided to publish it myself. Probably there are very few persons who would want to read the book anyway. The choice, or so it seems to me, is to prevent such persons from reading it by its price or by its lack of publicity. I prefer the latter.

This book is distributed by, and can be ordered from:
http://zeteticgleanings.com/
I welcome your questions, comments and feedback at:
info@zeteticgleanings.com

TABLE OF CONTENTS

CHAPTER I

THE HISTORICAL PROLEGOMENON

Section 1. The Challenge Problem.

Natural scientists and mathematicians are not different from other persons in their desire for the recognition of their achievements. To gain recognition requires the demonstration of originality - to copy someone else's work is obviously not sufficient. This is why the struggle to achieve priority has been for a long time the battleground on which the war for recognition has been waged. This is notoriously true for the natural sciences, as the histories of the search for the structure of DNA[1] and "The Hunting of the Quark"[2] in physics illustrate. It is, of course, equally true of mathematics. For example, John von Neumann tried to make it abundantly clear that Emile Borel deserved not even a particle of credit for the fundamental theorem of game theory.

In earlier times, at least in mathematics, the struggle took a different form. During the very long period of development of mathematical methods for problem solving it was common for mathematicians to propound difficult problems for their colleagues to attempt to solve. This was a straightforward way to demonstrate superiority: if you can't solve it and I can, then I am a better mathematician than you are or, at least, I am in command of better problem-solving methods than you are. Such, for example, was the challenge match between Fior and Tartaglia on February 22, 1535 concerning the solution of cubic equations. Each of the contestants proposed thirty problems to the other with the intention being that the one who could solve the greatest number within fifty days would be the winner. Tartaglia solved the problems proposed by Fior in two hours while Fior could not solve any of those proposed by Tartaglia. The same kind of contest occurred between Tartaglia and, as opponents, Cardan and his pupil Ferrari.

Subsequently, the emphasis changed from challenge matches between two opponents to challenge problems directed at any other interested mathematicians. A famous early example of this was the challenge problems addressed to all mathematicians by Pascal in 1658 with prizes offered for the first two solutions. No one won the prizes and when Pascal published his own solutions they created a sensation among mathematicians. Another such example was Pell's equation, proposed as a challenge problem in 1657 to all living mathematicians by Pierre Fermat.

The most famous examples of challenge problems, and the ones that led to their disappearance, were those initiated by Leibniz in 1687, all dealing with that area of mathematics known as the calculus of variations. Leibniz proposed the first two such problems as challenges specifically directed to the Cartesians. The first of these problems asked what curve would be such that a body falling due to gravity on that curve would cover equal perpendicular spaces in equal times. This is the well-known isochronous curve. The Cartesians were unable to solve the problem but Jakob and Johan Bernoulli, Huygens, and Leibniz all solved it. Leibniz also proposed a variation on this problem which leads to what is known as the paracentric isochronous curve. in 1690 Jakob Bernoulli proposed the problem that leads to the catenary curve and in 1696 Johann Bernoulli proposed the problem that leads to the brachistochrone.

In about 1695 the famous dispute between the English and the Continental mathematicians about the priority of Newton or Leibniz in the discovery of the calculus began to become heated. The challenge problems then began to be used as weapons in this fierce intellectual dispute. Thus, when David Gregory in 1697 published a detailed study of the catenary he was accused in a most unpleasant way of plagiarism by the Bernoullis. in 1716 Leibniz proposed a challenge problem to the English mathematicians that Johann Bernoulli had proposed to him in a letter of 1694. This is generally known as the problem of orthogonal trajectories. Newton, in fact, solved the problem the same night he heard of it but his published solution was more a general plan of attack than it was a detailed solution and Johann Bernoulli attacked it fiercely. Taylor defended it intemperately and Johann Bernoulli gave a bitter reply. Taylor sent a challenge problem but Johann solved it easily. John Keill then sent the last such problem to Johann: what is the path of a projectile in a medium that resists proportionally to the square of the velocity? Johann immediately solved this problem for any power of the velocity and it then developed that Keill was apparently unable to solve the problem. Johann trumpeted his triumph both crudely and loudly.

The bitterness produced by this series of challenge problems led to a general reluctance for this kind of contest and there has been hardly any other example except for the one shortly to be discussed. in retrospect, of course, it seems clear that Newton actually had priority in the discovery of the calculus but that Leibniz had independently developed at least parts of it and that the notation introduced by Leibniz was very much superior to that used by Newton. But because of the bitter dispute, English mathematicians continued to use the inferior notation for more than one hundred years. in the process, they fell very far behind the Continental mathematicians in terms of the quality of their work. Therefore, the English mathematicians had particular reason to reject the idea of challenge problems.

It is therefore quite surprising that George Boole, about 125 years later, saw fit to issue a challenge problem. While he did not use the words, "challenge problem". this is clearly what he had in mind and, furthermore, he was fully aware of what he was doing in the context of the history as outlined above. Since a version of his problem is of major importance in the present study, it is appropriate to quote Boole at length:

Proposed Question in the Theory of Probabilities

> Of those rigorous consequences of the first principles of the theory of probabilities the general utility of which has caused them to be ranked by Laplace among the great secondary principles of the science, none is more important than the following: - If an event E can only happen as the result of some one of certain conflicting causes A_1, A_2, ..., An, then if C_i represent the probability of A_i, and p_i the probability that if A_i happen E will happen, the total probability of the event E will be represented by the sum $\Sigma c_i\, p_i$.
>
> I am desirous of calling the attention of mathematicians to a question closely analogous to that of which the answer is conveyed in the above theorem; like it also, admitting of rigorous solution and susceptible of wide application. The question is the following: - If an event E can only happen as a consequence of some one or more of certain causes A_1, A_2, ..., A_n, and if generally C_i represent the probability of the cause A_i, and p_i the probability that if the cause A_i exist the event E will exist, then the series of values C_1, C_2, ..., C_n, P_1, P_2, ..., p_n, being given, required the probability of the event E.
>
> It is to be noted that in this question the quantity C_i represents the total probability of the existence of the cause A_i, not the probability of its exclusive existence; and p_i the total probability of the existence of the event E when A_i is known to exist, not the probability of E's existing as a consequence of A_i. By the cause A_i is indeed meant the

event A_i with which in a proportion p_i of the cases of its occurrence the event E has been associated.

The motives which have led me, after much consideration, to adopt with reference to this question a course unusual in the present day, and not upon slight grounds to be revived, are the following. First, I propose the question as a test of the sufficiency of received methods. Secondly, I anticipate that its discussion will in some measure add to our knowledge of an important branch of pure analysis. However, it is upon the former of these grounds alone that I desire to rest my apology.

While hoping that some may be found who, without departing from the line of their previous studies, may deem this question worthy of their attention, I wholly disclaim the notion of its being offered as a trial of personal skill or knowledge, but desire that it may be viewed solely with reference to those public and scientific ends for the sake of which alone it is proposed.[3]

The key difference between the problem given in the first paragraph of this quotation and the new one given by Boole is that the former assumes that the causes (A_i) are conflicting so that only one of them can occur. This entails the fact that the p_i of the first problem are the probabilities of E's resulting as a consequence of the corresponding A_i. In Boole's new problem several, or all, of the A_i could occur simultaneously. This forces a change in the conception of the p_i and this is why Boole carefully states that the p_i for his new problem represent the total probability of the existence of the event E when A_i is known to exist.

The problem as stated by Boole is certainly not perspicuous. in fact, while Boole discusses the problem as he formulated it in the quotation above , this general problem is not the one that was discussed at length by several persons and which is generally known as Boole's Challenge Problem.[4] The Challenge Problem is a special case of the stated problem. Specifically, suppose there are only two events, A_1 and A_2; suppose that E cannot occur unless at least one of the two events occurs; suppose that the probabilities of the two events are C_1 and C_2; and suppose, finally, that the conditional probabilities of E given either of the events are, respectively, P_1 and P_2. What, then, is the probability that E will occur?

It is not clear how this particular variant of the general problem arose. It first appears in print in an article by Cayley.[5] There Cayley says: "The following question was suggested to me, either by some of Professor Boole's memoirs on the subject of probability, or in conversation with him, I forget which; it seems to me a good instance of the class of questions to which it belongs." Since it does not appear in any of Boole's writings before the date of Cayley's article it would appear that the problem must have arisen in conversation. When Boole first refers to this specific problem, he says: "The question considered by Mr. Cayley in the Philosophical Magazine for October, 1853, p. 2559, is a particular case of a problem proposed by me in Problem is a special the Cambridge and Dublin Mathematical Journal in the month of November, 1851. This may justify my offering a few remarks upon the subject."[6]

Cayley suggests a possible instance of the problem. Suppose that if on a given day the wind speed is greater than w the day is called windy (A_1); suppose that if there is rain of amount greater than r the day is called rainy (A_2); suppose, further, that if the wind speed is greater than W or if the amount of rain is greater than R, or both, then the day is called stormy (E). Given the two p's and the two c's, what is the probability of E? Boole gives a different possible instance.[7] A field visible from the window of the room in which he works can be flooded (E) either as a result of floods from the upper sources of the river Lee (A_1) or as a result of tides from the sea (A_2) or, of course, as a result of both simultaneously. Given the p's and the c's, what is the probability that the field is flooded?

Here is a more modern version. A company is going to run an advertisement in two magazines (A_1 and A_2) that will invite the reader to request information concerning a new financial service. The request is E. in this version the c's would be the proportion of some specified target audience reached by the two magazines (called "reaches" in the industry) and the p's would be the proportions of the magazine readers expected to respond to such an advertisement. The desired probability of E translates here into the proportion of the target audience that will request the information. It will be seen that this is a "real-life" problem of some consequence in advertising allocation.

The obvious question is: Why did Boole think that this problem was sufficiently important to justify returning to the long-unused method of a challenge problem? To answer this question we need to discover what Boole had tried, and he thought successfully, to accomplish in the areas of logic and probability theory. This is the subject of the next Section.

Section 2. George Boole's Theory of Probability.

George Boole (1815-64) has the reputation, and moreover can afford the proofs, of being an outstanding thinker and mathematician. He was particularly unusual for his time in that he was essentially self-educated. He made a living by teaching at private schools, starting at age 16. When he was 25 he founded his own school. He studied mathematics in his spare time, began a correspondence with some Cambridge mathematicians, and began to publish. in 1844 he received a gold medal for his research in analysis. in spite of the fact that Boole did not have a university degree, in 1849 he was appointed Professor of Mathematics at the newly established Queen's College in Cork, Ireland. He wrote many articles and three important books: *An Investigation of the Laws of Thought, A Treatise on Differential Equations, and A Treatise of the Calculus of Finite Differences.*[8] Both of the last two of these became class-books at Cambridge and Reverend Harley said: "it is, we believe, unexampled, that works written by a self-taught man should attain to such a position in a university distinguished in the particular department of science to which they relate."[9] Boole died at a relatively young age as a result of walking two miles to his lecture in a downpour and then lecturing in his wet clothes. It is surely legitimate to note, although it is irrelevant to our subsequent discussion, that everything that is remembered about Boole contributes to a picture of a remarkably congenial human being.

It is certainly true that Boole's contemporary claim to fame rests entirely on his logical work. One would have to be virtually illiterate in mathematics not to have heard of Boolean algebra and its origin in the mathematical analysis of logic. Yet while Boole, if he came back to life, might be flattered by this he would undoubtedly find it both surprising and disappointing that there is hardly any interest in his theory of probability. Note the title and sub-title of his major work: "An Investigation of the Laws of Thought, On Which Are Founded the Mathematical Theories of Logic and Probabilities". It is clear from the title that Boole, at least, gave equal weight to both subjects.

In fact, persons who were much more nearly contemporary with Boole than we are thought that his major interest, or goal, was the theory of probability. Thus, Venn says: "It was indeed for the purpose of improving the calculus of Probabilities that Boole mainly worked out his system."[10] Charles Sanders Peirce says: "The principal use of Boole's Calculus of Logic lies in its application to problems concerning probability."[11] Peirce, again, says: "An essential, not to say vital, element of Boole's method lay in its applicability to the calculation of probabilities and statistical relations."[12] Boole himself says, after having given a number of examples of his logic: "in none of the above examples has it been my object to exhibit in any special manner the power of the method. That, I conceive, can only be fully displayed in connexion with the mathematical theory of probabilities."[13]

But regardless of the relative weights that Boole may have intended for logic, on the one hand, and probability theory, on the other, it is clear that he had a grandiose conception. in fact, it is doubtful that anyone other than Aristotle and Leibniz had a comparably encompassing conception. Boole intended nothing less than the development of a mathematical method, based on the "laws of thought", that would enable the answering of any well-formulated question in logic or probability. And Boole did not equivocate concerning what needed to be done. Here is his statement of the general logical problem: "Given a set of premises expressing relations among certain elements, whether things or propositions: required explicitly the whole relation consequent among any of those elements under any proposed conditions, and in any proposed form."[14] And here is his statement of the general problem in probability theory: "Given the probabilities of any events, simple or compound, conditioned or unconditioned: required the probability of any other event equally arbitrary in expression and conception."[15] Boole believed that he had solved both of these problems with the methods developed in his "An Investigation of the Laws of Thought".

In contrast to his own more general treatment of probability theory, Boole often refers to the accepted theory of probability at the time of his own efforts. With regard to this accepted theory, he says: "Various questions relating to causes and effects can be solved by known methods under the particular hypothesis that the causes are mutually exclusive, but apparently not otherwise."[16] This explains the importance Boole ascribed to the Challenge Problem described in Section 1. It is clear that in this problem nothing is known about the possible joint occurrence of the two causes and it is certainly not assumed that they are mutually exclusive. If the two causes were mutually exclusive then it is obvious that the probability of E would be given by $p_1c_1 + p_2c_2$. Since they are not mutually exclusive there is no way, using ordinary probability theory, of determining the probability of E. But Boole claims to be able to calculate exactly this probability. Therefore, the Challenge Problem is precisely an example where known methods are not sufficient and, Boole claims, his method is sufficient. This is why he considered the question important enough to justify re-introducing the old device of the challenge problem. And this is why Boole says, in the statement of the challenge problem quoted above: "First, I propose the question as a test of the sufficiency of received methods."

Needless to say, we will be considering the Challenge Problem at considerable length subsequently. For present purposes we may simply ask: What happened to it? As noted in Section 1, the English mathematician, Arthur Cayley, published a solution of it (1853) to which Boole took exception, claiming that it was not a solution to the problem as stated but, rather, to a special case of the problem. Boole offered his own solution (also in 1853) and at greater length in his book in 1854.[17] Henry Wilbraham (1854) strongly attacked Boole's solution and argued in favor of Cayley's. Richard Dedekind (1855) amplified Cayley's solution and argued in favor of it.[18] Cayley and Boole jointly discussed the problem further (1862) but inconclusively in the sense that each of them maintained his original position. Boole's death in 1864 ended the discussion without resolution. Some contributions subsequent to his death will be discussed below.

How might all of this be summarized for present purposes? I would put it that Boole had very bad luck with his Challenge Problem. First, he chose a poor example to illustrate his claim that his was a more general theory of probability. Innumerable better examples could easily have been given. I will give one immediately below. Second, he had bad luck in having Arthur Cayley as his opponent. Cayley was a very well-known mathematician who authored almost one thousand articles. Cayley's solution to the Challenge Problem as he originally presented it is anything but obvious but it did not require a Cayley to find it so perhaps someone else would have. However, when Cayley stated (in the interchange of 1862) that he did not understand Boole's argument it surely had a major negative impact on the reception given to Boole's theory. Third, Boole had bad luck in not paying close enough attention to Wilbraham's argument. Had he done so he would have been able to refute that argument much more clearly than he, in fact, did. Fourth, and finally, he had bad luck because his contemporaries, and posterity, took the Challenge Problem to be synonymous with Boole's theory of probability. Therefore, when it appeared that he had lost the argument about the solution to the Challenge Problem this was taken as a sufficient demonstration of the inadequacy of Boole's theory.

In fact, the treatment that history has accorded Boole's theory of probability is really quite extraordinary. Here is a man who is held almost universally in high esteem and who claims to have a theory of probability that enables one to solve very many problems that cannot be solved with the usual theory. Yet in the vast number of books dealing with probability theory or applications written in the past century one looks in vain for a reference to Boole. More specifically, John Maynard Keynes, in his book on probability theory, discusses Boole's theory (in 1921).[19] The next treatment, as far as I know, is the book by Theodore Hailperin (in 1976, with a second edition in 1986).[20] This is a remarkably long time to ignore a theory produced by a person of such intellectual stature as Boole.

This treatment would be justified either if the new kinds of problems that Boole claims to be able to solve are not of any interest or if the traditional theory is now able to solve them. Neither of these possibilities is true. As an example of an important problem of the kind Boole's theory can solve and the traditional theory cannot, consider advertising media data. It will be sufficient here to refer only to magazines but the same situations arise for any kinds of media (television, radio, newspapers, and so forth). Data can be purchased that give, for a specified target population, the proportion that will be exposed to a given specific magazine and the proportion that will be exposed to both of two specified magazines. Such data are available for large numbers of magazines and it is rarely the case that any similar data is available for combinations of greater than two magazines. in the industry such data are called reaches (for individual magazines) and duplications (for pairs of magazines). Suppose, then, that an advertiser is considering buying space in four magazines. The advertiser will know the reach of each of the four magazines and the duplication of each of the six combinations of two of the four magazines. in order to make economically rational decisions, the advertiser often needs to be able to estimate the proportions of the target audience who will be exposed to any three of the magazines or to all four of them. in the orthodox theory of probability there is simply no way to do this. in Boole's theory, on the other hand, it is a straightforward calculation. This kind of problem arises, in advertising, for any number of magazines or, more generally, for any number of media. It is clearly important and Boole claims to be able to calculate precisely the quantities that are needed. Therefore, neither of the two possibilities mentioned at the start of this paragraph is true. (This, incidentally, is an example of a type that Boole could have used as a challenge problem and that would not have been subject to the "bad luck" mentioned above.)

This strongly suggests that which is abundantly evidenced otherwise: Boole's theory of probability has hardly ever been discussed because it is generally considered to be wrong by those few persons who are familiar with it. There is surely good reason to consider how this came about.

Section 3. Appraisals of Boole.

Boole's theory of probability has as its foundation his theory of logic. But, generally speaking, Boole's contemporaries did not have a very high opinion of his logic. Styashkin discusses this.[21] Boole's contemporaries did not understand, or else they did not appreciate, his operation of addition. It will be recalled that for Boole x + y signified a class that excluded the common portion of the classes x and y. Thus, for Boole, x + x = 0. Jevons took particular exception to this and one of the major changes he introduced into Boole's logical system was the definition of addition inclusively.

More generally, most of Boole's contemporaries and immediate successors considered his logical system to be artificial and quite arbitrary. in fact, Jevons characterized Boole's system as an amazing combination of truth and mistakes. Most of the more traditional logicians were of the opinion expressed by Lotze to the effect that only the premisses and the conclusion were logical facts in Boole's system. Everything in between they thought was meaningless. There were even persons who maintained that Boole had by chance obtained valid results by false methods. This kind of objection was absurdly over-generalized but it did have a basis in Boole's presentation of his method. Thus, Boole says: "..., it is an unquestionable fact that the validity of a conclusion arrived at by any symbolical process of reasoning, does not depend upon our ability to interpret the formal results which have presented themselves in the different stages of the investigation."[22] This is Boole's defense of his method, which sometimes involved manipulations leading to expressions that were not subject to interpretation in terms of the original problem. Boole argued that this made no difference as long as the final equation could be interpreted correctly in terms of the original problem.

The history of logic is often divided roughly into three periods: from antiquity to 1847; from 1848 through the end of the nineteenth century; and from the beginning of the twentieth century to date. The reason for ending the first period at 1847 is that this was the year that saw the appearance of Boole's "A Mathematical Analysis of Logic". The first period is that of classical syllogistic logic, the second period is that of Boolean or algebraic logic, and the third period is that of mathematical logic. The extraordinary vigor with which a previously moribund subject was pursued in the second period is well shown by a count made by William Bartley of works on logic published in various time intervals:[23]

1798 - 1837	**4**
1838 - 1847	**0**
1848 - 1857	**3**
1858 - 1867	**8**
1868 - 1877	**31**
1878 - 1887	**100**

The last decade given includes the important works of John Neville Keynes, John Venn, and Lewis Carroll.

This remarkable efflorescence of logic during the period 1848-87 cannot be ascribed to Boole but it certainly involved his work in the sense that most of the workers in the field explicitly compared their efforts to those of Boole and explicitly tried to produce systems that would improve that of Boole. Some of the persons who made notable contributions during this period were Lewis Carroll, John Venn, Stanley Jevons, Alexander Macfarlane, Hugh MacColl, Charles Sanders Peirce, and the Johns Hopkins group associated with Peirce, of whom the best known was Mrs. Christine Ladd-Franklin. All of these persons were trying to remedy defects that they believed existed in Boole's logical system.

This ferment of activity ended as a result of the change associated with the work of Frege and, of course, the publication of Principia Mathematica, the start of the third period mentioned above.[24] This was unfortunate for Boole and his successors because it meant that their "revolution" was never really accomplished. As Bartley says: "Almost always essential to the success of a scientific revolution is the institutionalization of its doctrines in textbooks. But algebraic logic never quite reached the textbook stage."[25] The result was that the status of Boole's logical system was very much in doubt. Most of the contributors to algebraic logic had challenged, to a greater or lesser degree, Boole's theory. As of, say, 1900, a student would have had little reason to believe that Boole's logical theory was correct.

Of course, we now know that Boolean logic can be rigorously developed but even today it is a vexed question to decide exactly what arguments of Boole himself are correct. in other words, the logic that can be rigorously developed is likely to be called the Jevons-Schroder-Boole logic and sometimes Peano's name is also included. The logic is, in short, that which resulted from the modifications introduced by the other logicians. It requires careful analysis to determine how much of the mathematical rigor applies to Boole's original system.

In any case, it is surely a fair summary to conclude that, as of the latter part of the nineteenth century, it was generally believed that there was something seriously wrong with Boole's logic. Since Boole had very carefully erected his theory of probability on the foundation of his logic it would immediately follow that his probability theory would also be faulty. This opinion would be completely confirmed by the work of those algebraic logicians, mentioned above, who also considered probability theory. There were several of them. It seems to have been forgotten for many years but there is a very close relationship between algebraic logic and probability theory. As one person put it, algebraic logic ought to be simply the probability theory appropriate to situations where every event has probability either one or zero. Or as Clarence I. Lewis stated, in regard to an example of Boole's showing the use of his logic to resolve a question in probability theory: "It is a matter of surprise that this immediately useful application of symbolic logic has been so generally overlooked."[26]

Most of the explicit discussion of Boole's theory of probability dealt with the Challenge Problem but some others among his specific examples were also occasionally discussed. As would be expected, the only reason an author might discuss such an example was because he believed it to be wrong. Obviously, an author is only rarely going to devote space to a predecessor's examples if they are correct although the author may do so if he believes he has a better way to analyze the example. The net effect of the various claims and demonstrations of Boole's successors was to confirm the conclusion that Boole's probability theory was simply wrong.

In order to summarize the arguments made against Boole's theory of probability it is essential to note that there are two kinds of probability problems used as examples in "An Investigation of the Laws of Thought". One kind occurs in Chapter XVIII, "Elementary Illustrations". These are essentially examples that can be worked, often with less difficulty, by standard methods. The other kind occurs in Chapter XX, "Problems on Causes". These are all problems similar to the Challenge Problems, which is also included in this Chapter, and they share the characteristic that they cannot be solved by the usual methods of probabilistic analysis but they can be solved by Boole's methods.

A fair summary of the kinds of conclusions reached about problems of the first type might be that it seems clear that Boole made some mistakes but that none of them are particularly serious. An instance of this would be Boole's Example 3 of Chapter XVIII which deals with a simple case of the combination of testimony.[27] Boole's solution is correct but he states: "It is probable, from the simplicity of the results

(5) and (6) that they might easily be deduced by the application of known principles; but it is to be remarked that they do not fall directly within the scope of known methods. The number of the data exceeds that of the simple events which they involve." Alexander Macfarlane discusses this example and, quite correctly, points out that Boole's quoted statement is wrong since it is perfectly straightforward to solve the problem by standard methods.[28] in fact, it is a good deal easier than the solution using Boole's method. Again, Charles Sanders Peirce considered Boole's Example 6.[29] From this example Boole concluded: "... propositions which, when true, are equivalent, are not necessarily equivalent when regarded only as probable." Peirce says this is absurd and further states that Boole had incorrectly formulated the problem and then had incorrectly solved the wrongly formulated problem. It is certainly true that this particular example is of no consequence for the truth or falsity of Boole's theory of probability. But whether Peirce is right or not may well never be known because of the notorious and almost insuperable difficulties of understanding Peirce. It is surely worthy of mention that Hailperin writes that "The solution that Peirce gives, involving seven introduced parameters, is incomprehensible to us." (Hailperin: Boole's Logic and Probability; p. 284)

The only problem of the second type (Chapter XX) discussed in this literature is the Challenge Problem. It will be recalled that Cayley, Wilbraham, and Dedekind all disagreed with Boole's solution of this problem. Macfarlane and MacColl each discussed this problem and each concluded that Boole's solution was wrong.[30] Each independently reached the same conclusion that Wilbraham had reached: the data are insufficient to permit an exact solution. The most that could be accomplished would be to determine bounds between which the desired probability would fall. Perhaps indicative of the importance ascribed to the Challenge Problem may be that Macfarlane ends his book with a comment on Boole's solution: "What is given by Boole's solution is not the mathematical probability of the event E, but the most probable value of the probability which can be deduced from the given data." There will be further comment on MacColl's analysis of the problem in the next Section.

Suppose someone had wished to attempt to defend Boole's theory of probability against this onslaught. Such a person would have had to cope with the very considerable difficulty and unpleasantness of the required analysis. Boole's method, as he presents it, is opaque, onerous, and disagreeable. Concerning this, Keynes says: "... the difficulty of his method is so great that I do not know of any one but himself who has ever attempted to use it."[31] And here is Boole himself, towards the end of the examples of Chapter XX: "For this cause it has not been deemed superfluous to exhibit in this chapter its application to problems, some of which may possibly be regarded as repulsive, from their difficulty, without being recommended by any prospect of immediate utility."[32] And, indeed, it is very easy to regard some of Boole's analyses of Chapter XX as repulsive because of their difficulty.

With all of the arguments and evidence it seems fair to summarize that by the latter part of the nineteenth century the jury had returned with its verdict on Boole's probability theory: not true. It remained, so to speak, for the judge to pronounce a sentence. This was done by John Maynard Keynes, who deserves a separate Section.

Section 4. John Maynard Keynes.

Had John Maynard Keynes never written a book on economics he would still deserve to be famous because of his great book on probability theory. This book appeared in 1921 when Keynes was 38 years old. Keynes states in his Introduction to his Bibliography that while the bibliography does not pretend to be complete, it has "a much longer list of what has been written on probability than can be found elsewhere."[33] Later he says: "I have not read all of these books myself, but I have read more of them than it would be good for any one to read again."[34] It is certainly true that even a person quite thoroughly immersed in the older literature of probability theory will rarely find a contribution of consequence that Keynes does not discuss.

Probability theory is, of course, of major importance in every field of knowledge and even in philosophy. Despite this, as Keynes points out: "It is now fifty-five years since Dr. Venn, still an accustomed figure in the streets and courts of Cambridge, first published his "Logic of Chance"; yet amongst systematic works in the English language on the logical foundations of Probability my Treatise is next to his in chronological order."[35] Because of this combination of circumstances and because Keynes's Treatise is a work of such exceptional scholarship, it was immediately recognized as a classic.

The emphasis today among both students and teachers of probability theory is almost entirely on its mathematical side. This was really initiated by the appearance in 1950 of another great classic, William Feller's "An Introduction to Probability Theory and Its Applications".[36] Both sides are needed but currently Keynes seems to have been more nearly forgotten. Feller says, in discussing the Petersburg paradox: "... but the modern student will hardly understand the mysterious discussions of this `paradox'."[37] One might perhaps better say that unless the modern student understands, for example, Keynes's insightful discussion of the Petersburg paradox the student will be very poorly prepared to use his or her mathematics.

In any case, for very good reasons Keynes's Treatise was generally considered to be a reliable summary of the state of knowledge about probability theory as of 1920. This means, in particular, that his comments about Boole's theory of probability were generally taken to be the definitive statement on the subject. This was very bad luck for George Boole because Keynes truly administered the coup de grace to Boole's system. Keynes certainly thought highly of Boole's abilities. Commenting on one of Boole's memoirs, Keynes says: "..., the memoir is as full as are his other writings of genius and originality."[38] However, Keynes believed that Boole had made a fundamental error that vitiated much of his analysis.

The first reasonably clear statement of this presumed error was made by Hugh MacColl.[39] It will be recalled from our earlier discussion that precisely the reason why Boole proposed his challenge problem was that the desired probability could not be calculated by the usual theory and it could be calculated by Boole's theory. All of the persons who commented on the problem originally - Cayley, Wilbraham, Macfarlane, MacColl - were perfectly capable of the simple task of counting equations and unknowns to reach the conclusion that the Challenge Problem was not solvable by the usual theory. This, of course, was exactly Boole's point. However, no one of these persons seemed able or willing to undertake the study of Boole's theory to discover how and why he succeeded in solving the problem. They cannot be faulted for this because of the great difficulty of understanding Boole's theory. This granted, two obvious approaches are both exemplified by MacColl. First, one can deduce by the usual theory bounds for the desired probability. MacColl did this and claimed that Boole's solution must be wrong because it violated the bounds.[40] This claim, incidentally, is incorrect.[41] Second, one can attempt to find the error that Boole "must" have made. It does not require much searching to find a suitable candidate since Boole ascribes great importance to the assumption that two variables are independent unless the contrary is known. MacColl took this step and claimed "that Boole's fallacy turned on his definition of independence."[42]

Keynes quotes MacColl approvingly and gives him credit for having been the first to recognize the source of Boole's errors. Keynes adds that Wilbraham had pointed out the error "accurately though somewhat obscurely". Keynes himself sharpens the point considerably by stating: "The central error in his system of probability arises out of his giving two inconsistent definitions of `independence'."[43] Keynes claims that this error of Boole's makes his solutions of his Examples I - VI of Chapter XX all incorrect.[44]

A reader at this point would be pardoned for thinking that Boole's theory of probability is a total loss. Such a reader would simply be recapitulating the consensus of informed opinion about this theory since, after all, this consensus was based on the same evidence reviewed here. However, such a conclusion is, and was, quite wrong.

It is time to redress the balance and begin to make a case for Boole's theory of probability. I would like to be as charitable as possible to Keynes but it is impossible to be fair to Boole and his theory without stating categorically that Keynes provides no evidence of the slightest understanding of Boole's theory. I very much doubt that he ever was able to solve any one of Boole's examples, even the elementary illustrations, using Boole's method. It is easy to construct a scenario that might explain this. On the one hand, Keynes had assimilated an enormous literature of fair difficulty on probability theory. On the other hand, he was anxious to present to the world his own extremely innovative theory. It will be recalled that it is in the Treatise that Keynes introduces his symbol a/h, where h is the premiss of an argument and a is its conclusion. in Part II of the Treatise he brilliantly demonstrates the rules for using this symbol - essentially, the ordinary conditional probability - and shows that the usual rules of inference follow as theorems from his definitions. Further, he then uses his symbol and these rules to solve many problems in probability theory, including the Challenge Problem and other examples from Boole's Chapter XX. At the time of publication he was a relatively young man with many urgently competing interests in economics. Anyone who has attempted to overcome the great difficulties barring the way to an understanding of Boole's theory will surely sympathize with Keynes when, apparently, he searched for a reason to dismiss Boole's theory rather than attempting to understand it. The alternative, even for a person of Keynes's ability, would be some months of hard intellectual effort with the corollary delay in the publication of his book and in all of his other activities.

This scenario is, of course, hypothetical. However, it is categorical that Keynes did not demonstrate any understanding of Boole's theory. A reasonable appraisal might be that MacColl's conclusion that Boole had erred in his definition of independence would have been particularly congenial to Keynes because in his Treatise Keynes puts considerable emphasis on this kind of mistake.[45] He might, then, have concluded that it was unnecessary to undertake the heavy effort to disentangle Boole's often difficult exposition because, after all, Boole's theory was built on a faulty foundation. in any case, Keynes did sharpen the criticism. Here is his statement at some length:

> "The central error in his system of probability arises out of his giving two inconsistent definitions of `independence'. He first wins the reader's acquiescence by giving a perfectly correct definition: `Two events are said to be independent when the probability of the happening of either of them is unaffected by our expectation of the occurrence or failure of the other.' But a moment later he interprets the term in quite a different sense; for, according to Boole's second definition, we must regard the events as independent unless we are told either that they must concur or that they cannot concur. That is to say, they are independent unless we know for certain that there is, in fact, an invariable connection between them. `The simple events x, y, z, will be said to be conditioned when they are not free to occur in every possible combination; in other words, when some compound event depending upon them is precluded from occurring. ...Simple unconditioned event are by definition independent.' in fact, as long as xz is possible, x

> and z are independent. This is plainly inconsistent with Boole's first definition, with which he makes no attempt to reconcile it. The consequences of his employing the term independence in a double sense are far-reaching. For he uses a method of reduction which is only valid when the arguments to which it is applied are independent in the first sense, and assumes that it is valid if they are independent in the second sense. While his theorems are true if all the propositions or events involved are independent in the first sense, they are not true, as he supposes them to be, if the events are independent only in the second sense."[46]

What is to be said about this argument of Keynes? Two statements are sufficient. First, it is false in the sense that it is based on a misreading of Boole. When Boole is speaking of conditioned events, admittedly somewhat obscurely, he is not talking about ordinary conditional probabilities. "Conditioned" for Boole means that some logical descriptive equations are required in formulating the problem. Keynes apparently reads Boole to the effect that, to use Keynes's example from the previous quotation, if xz is possible then x and z are independent. This, of course, would be ridiculous. Let us state it precisely. If the probability of xz is known then it is part of the data of the problem. p(xz) can be any number and can bear any relationship to p(x) and/or p(z) that is not prohibited by the usual arithmetic of probabilities. If p(xz) is not known then xz may be conditioned in that the joint event either cannot occur or must occur. If p(xz) is not known and xz is not conditioned then it is assumed that x and z are independent. This is obviously completely different from Keynes's reading. Even a reading of the Challenge Problem should have shown Keynes his error. Second, Keynes's criticism is irrelevant. By his own statement, Boole's theorems are true if the events are independent in the first sense. Surely, then, Keynes should have attempted to indicate what kinds of problems might fall under the first sense. Had he done so I believe he would have discovered that almost all of the problems Boole deals with are in the first sense. in any case, it is surely true that major areas to which I maintain Boole's methods are currently applicable are of the kind Keynes refers to as the "first sense".

Further, Keynes makes no effort to show how either of the definitions would affect Boole's argument. in fact, nothing in Boole's theory needs, or in any way depends upon, any definition of independence more profound than: two events are independent if and only if the joint probability of the two events is calculated by multiplying the probabilities of the two events taken individually (in symbols, p(A,B) = p(A)*p(B)). Had Keynes been at all familiar with Boole's procedures he could not possibly have believed that there was any problem whatsoever with the definition of independence. Further, in this context Keynes's comment on Cayley is quite interesting. He references Cayley's solution (1853), Boole's answer claiming the solution was incorrect (1854), and Dedekind's defense of Cayley (1855). Keynes then comments: "The difference arises out of the extreme ambiguity as to the meaning of the terms as employed by Cayley."[47] This comment is really not to the point at all and "ambiguity" is a dubious word to use in this context. However, the interesting thing is that one of the few important articles not included in Keynes's bibliography is the article by Cayley, but almost jointly with Boole, that appeared in 1862.[48] in this article Cayley makes completely clear how his solution is obtained. Had Keynes read this article he would have seen that Cayley, using generally accepted arguments, was able to solve the Challenge Problem and he would have had a quite clear statement by Boole as to the reason for the differences between the two approaches. It would have been impossible for Keynes to sustain his thesis about the cause of Boole's "error" had he been aware of this reference.

In this same context it is also worth noting that Keynes does not discuss any of Boole's elementary examples (Chapter XVIII), with one exception that is not relevant to the present purpose. Since Boole's method worked perfectly well on simple examples that could be solved by orthodox methods and since Boole usually got the same answer it becomes more difficult to argue that he has made some fundamental mistake. These comments are not intended in the slightest degree to impugn Keynes's motives or scholarship. Like most of us, he perhaps had a tendency to perceive the world in a way that would be in accord with his theories but if this were a crime, who would escape whipping? in any case, it is essential for our purpose to emphasize that the claim that Boole somehow made an important mistake in his definition of independence is totally without foundation.

Quite apart from this misplaced basic criticism, Keynes shows in other ways a complete lack of insight into Boole's system. Thus, Boole attempted to show that Cayley's solution was wrong by taking particular values of the given probabilities which meet the requirements of the problem as stated and for which Cayley's solution is obviously incorrect.[49] Keynes tries to do the same thing for Boole's solution.[50] However, Keynes does this by assuming that A1 and A2 cannot both occur. The scantiest knowledge of Boole's system is sufficient to demonstrate that this assumption would require that the problem be formulated differently. (Incidentally, it is an example of what Boole meant by "conditioning".) It is easy to formulate it correctly and then, of course, the answer obtained by Boole's method is precisely the one Keynes gives. It is worthy of note that immediately after giving this erroneous counter-example, Keynes says: "The mistake which Boole has made is the one general to his system, ...".[51]

In summary, then, Keynes rejected Boole's system for invalid reasons. Undoubtedly, Keynes's basic motive was the same as that of his various predecessors who disputed Boole's claims. All of them essentially rejected the possibility that Boole could, in fact, calculate exact probabilities in cases with insufficient data to permit them to be calculated by the orthodox theory. The Challenge Problem, as I have stated several times, is of exactly this kind. Macfarlane, MacColl, and Keynes all argued that the best that could be accomplished was to determine bounds within which the desired probability would fall. Keynes used his own new stroke function, a/h, with incomparable facility to determine these bounds. He repeatedly states that the bounds so obtained are the answers, and the only possible answers, to the various problems of this kind that Boole had propounded.

The prestige of Keynes was so great, and the quality of his Treatise was so fine, that his conclusions regarding Boole's system were generally accepted. The Treatise was, indeed, the coup de grace for Boole's theory of probability. It was 55 years before there was another major discussion of Boole's system. Before we devote a section to this next major discussion, by Professor Theodore Hailperin, let us consider the predicament of any prospective student of Boole's probability theory during the long 55 years between Keynes and Hailperin.

Section 5. Studying Boole's Probability Theory.

We will subsequently see some of the many reasons why persons working on data in the form of contingency tables might be led to take an interest in Boole's probability theory. But even without having such potential applications of his theory, surely it is not unreasonable to expect that persons with an inclination towards mathematics might wonder what a person with George Boole's splendid reputation might have done in probability theory. Any person with an awareness of George Boole's known accomplishments would certainly be intrigued by the resounding title of his major work: *An Investigation of the Laws of Thought*, On Which Are Founded the Mathematical Theories of Logic and Probabilities (or LT for short). Really, how could someone avoid wondering about the substance underlying such a cymbal-crashing title?

Suppose, then, that this person with a new-found interest in Boole's probability theory has discovered this interest during the long 55 years during which there were no new attempts to explain Boole's argument.

So, a typical progress would then be to buy or borrow LT and start studying it. So doing, the potential enthusiast would quickly discover that everywhere in LT the text and examples hint at fascinating possibilities - and constantly emphasize that these possibilities are not available in the standard theories. But even the simpler of the examples turn out to be very difficult to duplicate and the algebraic burden is almost insurmountable. This is so much the case that it is easy to conclude that one really does not understand anything Boole is talking about. So, our prospective student - if he or she has not already previously done so - will search for an exposition of Boole's theories, both of logic and of probability theory. The student will not find any expositions but he or she will certainly find some of the negative assessments of Boole's theories that we have summarized above. There will be, in short, no evidence that anyone thinks, or thought, that Boole's probability theory might be correct and there will be strong evidence that the theory is very difficult. In this latter regard, recall the quote previously cited from Keynes:

> ". . . the difficulty of his method is so great that I do not know of any one but himself who has ever attempted to use it."

Perhaps two of Boole's easier probability examples may serve to illustrate the comments of the preceding paragraph. Here are Boole's final equations of his first probability example in Chapter XVII of LT, entitled "Demonstration of a General Method for the Solution of Problems in the Theory of Probabilities":

$$\frac{p'}{p'q' + p'(1-q') + q'(1-p')} = p, \quad \frac{q'}{p'q' + p'(1-q') + q'(1-p')} = q$$

These are two simultaneous equations and we are to solve for p' and q'. They are annoyingly difficult to solve but it is easy to verify that Boole's answer is correct:

$$p' = \frac{p+q-1}{q} \quad \textit{and} \quad q' = \frac{p+q-1}{p}$$

Here is the end of Boole's second example in Chapter XVIII of LT, entitled "Elementary Illustrations of the General Method in Probabilities": he obtains two simultaneous equations in s and t and, using Boole's notation, an equation showing the desired answer, a probability, as a function of s and t:

$$\frac{s}{p} = \frac{t}{q} = s + (1-s) \quad \textit{and} \quad \textit{Prob. } w = \frac{st}{s + (1-s)t}$$

Boole wants the answer for the probability to be in terms of p and q. That means one must solve the two equations on the left for s and t in terms of p and q and substitute in the expression for the probability of w. It is annoyingly lengthy to do so. The general rule is: even the seemingly easy steps in Boole are likely to be difficult. And these are two of the easiest examples of Boole's probability theory.

On the other hand, the great majority of his examples are very difficult ones. Boole was anxious to display the power of his methods and a good way to do so was to choose complicated examples and then show that his methods could still produce the correct answer. Thus, his Example 5 in Chapter IX, "On Certain Methods of Abbreviation" deals with five properties (A, B, C, D, E) associated with an unspecified class of "natural productions". There are three premises and each one involves fairly complicated relationships among all five of the properties. One of the questions he wants to answer for this example is, if E is eliminated from consideration, what does the presence of property A tell us with reference to properties B, C, and D? I made many false analyses before I succeeded in duplicating Boole's conclusion.

So, during the 55 years between Keynes and Hailperin there were many difficulties in the way of a student who wanted to gain an understanding of Boole's theory of probability (and, for that matter, the same applies to Boole's logical theory). First, the methods themselves were of great difficulty - recall the Keynes' quotation - and Boole himself referred to the difficulties of application being sometimes "repulsive". Second, the great majority of the examples Boole selected were of considerable complexity. Third, Boole's exposition of his methods leaves a lot to be desired. Fourth, and of primary psychological importance for the would-be student, with almost perfect unanimity those authorities who commented at all on Boole's theories considered them to be wrong.

Perhaps the important question is why significant numbers of would-be students persevered in spite of the obstacles listed in the preceding paragraph. Reporting only on the basis of my experience plus that of a significant number of friends and colleagues there seem to be three major reasons. First, the negative comments about Boole's theories amounted to "over-kill". Put simply, it was not easy to accept that a person of Boole's attainments could work for years on his major intellectual project and still be so grossly in error as his critics maintained. Second, it was an indubitable fact that Boole often applied his probability theory to problems to which the orthodox theory could provide answers. Boole's answers, in such cases, were always in accord with those of the orthodox theory. It would require a miraculous coincidence for this to be true while simultaneously Boole's theory was nonetheless false. Third, a major objection to Boole's theory was that it often produced unique answers when the orthodox theory could only provide bounds on the answer. One could not help noticing that Boole's answers were always in good accord with the bounds provided by the orthodox theory. This immediately suggested that Boole's theory might be extremely useful for the many real-life probability problems for which bounds on the answer are not sufficient: a unique answer is needed.

This was the intellectual situation during the long interval between the magisterial book by Keynes and the extremely fine book by Professor Hailperin. We must now consider how Professor Hailperin's book changed the scene.

Section 6. Theodore Hailperin

It was not until 1976 that George Boole's theories of logic and probability were again the subject of a major study. In that year he finally received what he had long deserved, a book-length study. This, of course, was the very fine book by Theodore Hailperin, Boole's Logic and Probability; North-Holland (Amsterdam). A considerably expanded second edition appeared in 1986.

This is a beautiful book with virtually endless merits. I cannot recall another mathematical book from which I received so much pleasure as in reading Hailperin's superb work. I would like to offer some brief comments about three major aspects of the book that are very much relevant to our purposes.

First, Hailperin had to find a much solider basis for Boole's arguments than Boole, limited by the mathematical knowledge of his time, found himself. It needs to be remembered that opinions of the "authorities" were often scandalously negative concerning both of Boole's theories. Boole's pupil, Stanley Jevons, characterized his teacher's system as an amazing combination of truths and mistakes.[52] Lotze was of the opinion that, concerning Boole's results, only the premises and the final equation were logical facts and that the intervening procedure was a meaningless paradox.[53]

Thus, as a rigorous basis for Boole's logic, Hailperin found that correct models are obtained "if we consider, not classes, but multisets as the entities over which the variables range."[54] For Boole's General Problem in probability Hailperin constructed a solution using linear programming. There is no doubt that these new bases for Boole's theories do accomplish their desired purpose. But, second, throughout Hailperin's book we find numerous statements such as the following sample:

"Clever as they were, the reasons Boole gave for his rules of interpretation convinced no one. In § 2.7 we show not only how to justify Boole's procedure here but to make sense of it in all respects."[55]

"From the modern point of view Boole's doctrine of primary and secondary propositions has serious conceptual flaws but these need not concern us as there were no deleterious consequences."[56]

"Our result here will also show that, for class calculus problems expressible in the form of a Boolean equation for an unknown, Boole's method always gives the correct solution, as the large number of worked examples in the Laws of Thought bear witness."[57]

"We now show that Boole's method always gives the correct result for Boolean algebra problems."[58]

"in § 3.3 we take up the L.T. approach, i.e. the one based on the notion of portions of time and, in § 3.4 explain why it "works" even though there is some semantic confusion."[59]

The uniform theme of these quotations is that however rickety may have been the structures that Boole erected for his theories, his conclusions were correct in terms of the more solid foundations subsequently provided by contemporary mathematical rigor and, in particular, provided by Professor Theodore Hailperin in his book. As one of the implications of these quotations, note that the would-be students of Boole's theories, discussed above, were correct to deduce that since Boole's methods gave correct answers for known problems it followed that the same methods might be giving correct answers for previously unknown problems. And note that J. M. Keynes could have spared himself some false statements about Boole's theories if he had followed the same line of reasoning. A major part of the reason for this outcome was that Boole was often considered to have an exceptional "mathematical intuition" that almost always protected him from errors that might well have appeared reasonable to other mathematicians. For one mathematician's views of Boole's intuition, see the footnote.[60]

Generally, Professor Hailperin's book has many summary statements that provide a fine historical view of the developments that affected Boole's theories. Here are a few:

"Never clearly understood, and considered anyhow to be wrong, Boole's ideas on probability were simply by-passed by the history of the subject, which developed along other lines."[61]

"Accordingly, he (Boole) views the general object of a theory of probabilities as: 'Given the probabilities of any events, of whatever kind, to find the probability of some other event connected with them."[62]

(On Boole's probability theory): "Here too, as with his logical system, we shall find that despite erroneous ideas and unnecessarily complicated methods there is much of interest and value to be gleaned."[63]

Finally, Professor Hailperin identifies and credits to Boole the major contributions for which he was responsible. Here are some examples:

"Although he makes no use of this method for finding narrowest bounds it is nevertheless of particular historical interest, for what Boole has here, clearly formulated, is a linear programming problem."[64]

"Our discussion of the problem of absolute probabilities is long and tedious, but we felt it worthy of attention as it is a key feature of Boole's method, a proof of which (in the general case) baffled him a long time. Involved is a rare example of a mathematical theorem which gives necessary and sufficient conditions for a rather general system of algebraic equations to have a unique solution."[65]

"That he formulated the problem of finding the narrowest bounds as a linear programming problem, in both dual and primal forms, and used Fourier elimination to find the bounds in the latter case, should be counted among his interesting and noteworthy achievements."[66]

I hope that my great admiration for Professor Hailperin's very fine book has been suitably expressed. If it has been, then the question that may be (perhaps, ought to be) in the present reader's mind will be to the effect, "Since Hailperin's book is so excellent, why do we need another one?" This is an important question and a major purpose of this Chapter is to answer it.

In order to do so we need to briefly review the major strand of criticism of Boole's probability theory by all those "experts" who were so convinced that his theory was hopelessly wrong. We will subsequently discuss this criticism at length but a brief summary will be sufficient for our present purpose. The major strand of criticism had to do with the relationship between the number of unknowns in a probability problem and the number of known equations.

Consider a simple probability problem involving the occurrence or non-occurrence of two events, A and B, where the non-occurrence of A is shown by A' and the non-occurrence of B by B':

	A	A'
B	a	b
B'	c	d

We are given the probability that event A occurs, P(A) = p, and the probability that event B occurs, P(B) = q. The small letters within the table are the probabilities that both the row and the column events occurred. Thus, a = probability that A occurred and B occurred, c = probability that A occurred and B did not occur. A' means not-A occurs, otherwise phrased as A did not occur, and similarly for B'. Now, we can write equations, using the small letters, for the pieces of information we are given: we know that

$a + b = q$

$a + c = p$

and

$a + b + c + d = 1$, since one of the small letters must occur.

We have, then, three equations and there are four variables. We can conveniently write our total information thus:

$b = q - a$

$c = p - a$

$d = 1 + a - p - q$.

The orthodox theory of probability cannot fill out the contingency table - meaning, find values for a, b, c, and d - unless there are as many equations as there are unknowns (the small letter variables). Here there are four unknowns and only three equations. Therefore, orthodox probability theory cannot answer any questions about the values of the small letters. Examples of such questions are: What is the probability that both events A and B occur (meaning, what is the value of "a"); What is the probability that neither event A nor event B occurs (meaning, what is the value of "d"); or, What is the probability that only one event occurs (meaning, what is the sum of "b" and "c").

Boole's theory of probability, on the other hand, can easily solve this problem. From Boole's theory we find that ad = bc, whence it follows with a little algebra that a = qp. This, then, sets the stage for what we have referred to as "the major strand of criticism" of Boole's theory of probability. It doesn't require much knowledge of mathematics to know that generally one needs as many equations as there are unknowns in order to solve for the unknowns. This is what is required by the orthodox theory and it is not true for our simple problem. Therefore, the orthodox theory certainly could not solve for the four unknowns: a, b, c, and d. But, argued the orthodox critics of Boole, the same limitation certainly applies to Boole's theory. If he obtains values for the four unknowns then he has, surreptitiously or unknowingly, introduced a fourth equation and unless he can offer an adequate argument as to where that fourth equation came from, his "solution" is invalid. And, of course, the critics maintained that no such valid argument could be found and, therefore, Boole's theory was wrong.

The orthodox critics said - in general but also for a problem such as our example - that the best that could be done was to develop bounds for the solution. It is quite easy to find bounds for a solution of our example. Let us work with specific numerical values. Suppose q = .6 and p = .45. Suppose, further, that we want bounds on "a". Consider a minimum bound for "a". We cannot have negative probabilities so $a \geq 0$ but from the requirement that $d = 1 + a - .6 - .45 = a - .05 \geq 0$ we have $a \geq .05$. This is the overall minimum possible value for "a". None of the four entries in our contingency table can be negative so it follows that the maximum possible value for "a" is $a \leq .45$, from the requirement on "c". Let us compare the resulting solutions for the four probabilities:[67]

	Boole	Min "a"	Max "a"	a = .09
a	.27	.05	.45	.09
b	.33	.55	.15	.51
c	.18	.40	0	.36
d	.22	0	.40	.04

As we will see, there are many real-world probability problems for which a specific answer is needed - not just bounds on a solution. It can be seen immediately that here - as is very often the case - the bounds are quite far apart and it is not at all clear how such bounds could be useful in many decision problems. The orthodox critics, of course, would respond that the bounds are the best that can be done with the existing information and that there is no use complaining about it. But that is perhaps the point: Boole claimed to find an exact answer. If his claim can be justified then his solution is, indeed, available as an alternative to the wide bounds from the orthodox theory. Further, it is the case that there are a fair number of other arguments that produce specific values in the analysis of such problems. What are the virtues and deficiencies of these arguments? How do they compare with Boole's theory?

Notice that Boole's exact answer for "a" is in the middle of the bounds - actually, just a bit more than the mean of the minimum and the maximum. This, however, is mostly a happenstance of our example. Suppose we ask, Is there an argument that leads from maximum and minimum bounds to a unique answer? The answer is that there is such an argument and that it is due to G. Polya.[68] We will present his argument later. For our present discussion it is sufficient to note that Polya's unique answer for our example is a = .09, which is quite far from Boole's answer of a = .27.[69] This difference is much more typical than our mentioning that Boole's value for "a" is close to the mean of the minimum and maximum values.

We are now ready to return to the question, How is the present book to be justified, granted that Hailperin's superb book is already available? The overall answer is that Professor Hailperin fully accepted the contention of the orthodox critics that Boole's method introduced equations that were not part of the data and that his "solutions", therefore, were of a different problem than the actual problem with a lesser number of equations. As the capstone of his very fine book, Professor Hailperin offers his own analysis of how to solve Boole's general probability problem. He writes, "in this section we are going to show that Boole's general problem can be solved without additional assumptions, solved not in his sense but in the sense of giving best possible upper and lower bounds on the probability of the objective event, these bounds being determined by the given data. Conceivably such a solution could have been presented by Boole for, as mentioned in § 4.7, the idea of finding bounds on the probability of an event originated with Boole, as well as the idea of obtaining, in his phrase, 'the narrowest limits'. Applying this technique of finding narrowest limits for the probability of the objective event would then be solving the general problem as we view it. But as Boole never conceived of this as a solution to the general problem it is left for us to present it this way."[70] This is the area in which Hailperin showed that obtaining such bounds was a reasonably straightforward linear programming problem. He presented a rigorous development of the linear programming solution so that it could be applied to arbitrary problems that Boole took as examples of his solution of the "general problem of probabilities".

In fact, Hailperin outdid the orthodox critics of Boole's theory by a considerable margin. First, he demonstrates that Boole's general approach can be rigorously justified. Then he argues that what orthodox probability theory could do when there were insufficient equations to support specific values for all of the unknowns, the determination of bounds on the unknown probability, was the orthodox solution to the problem. And, finally, he demonstrated how to find the best bounds, using linear programming. One could hardly ask for more!

Still, this leaves open all questions about Boole's theory of probability. Does it have any intellectual merit? How does Boole obtain the answers he offers to his examples? How does Boole's theory compare to the several other contemporary methods of finding a specific answer when there are insufficient equations to support a general solution? in defending his method, Boole maintained, in answer to the orthodox critics, that any equations that were added were simply the result of the operation of the basic laws of thought. Is that true?

All such questions about Boole's theory of probability remain to be answered. One can generalize slightly. There are many probabilistic decision problems for which one needs a specific answer for one or another related probability. Bounds on the probability might be helpful if they were reasonably narrow bounds but they often are not. Boole's theory is one that provides a specific answer but there are others. How do they compare? Which is the most appropriate one to use in such cases? The present book is my attempt to answer these questions.

Section 7. The Plan of Attack.

Since Boole explicitly developed his theory of probabilities on the basis of his theory of logic we will begin by summarizing his logic to the degree necessary for his probability theory. This summary may not be adequate for understanding Boolean logic starting from "scratch", so to speak. But most persons have encountered the basic ideas of Boolean logic in one or another course and our summary should be sufficient to refresh their knowledge to a degree adequate for our purposes.

Then we will present Boole's method in probabilities as he used it himself. in so doing, we will find that there are some simple changes that permit the necessary calculations to be done very much more easily. These changes do not affect the underlying method, it is still entirely Boole's. I believe that these changes also make it much simpler to understand what assumptions underlie Boole's method and I will try to state the necessary assumptions. All of this, including these simplifications, will be presented and illustrated in Chapter II, using various examples that Boole himself used.

The whole of Chapter III will be devoted to the Challenge Problem. The various disagreements can be very much better understood in terms of the simplified procedure and we will be able to see exactly what differences existed between Boole and Cayley. Still using only Boole's own theory, albeit very much simplified, we will see, for example, that Boole could quite easily have made a much more impressive rejoinder to Cayley, particularly in the 1862 article.[71]

However, when we examine Wilbraham's comments we will find that there definitely seems to be an error in Boole's general method.[72] in fact, Wilbraham's article is not very perspicuous and the essential point is clearly made in another, very short, article by Cayley that is more or less a postscript to the 1862 article in collaboration with Boole.[73] Whether it is really an error or not is almost a philosophical question. My own opinion is that it is not an error and I am quite certain that that was Boole's opinion. The question will be discussed in Chapters III and IV. In any case, the cause of this possible error can be exactly identified and it is so simple that it can be stated here although the argument will be postponed until Chapter IV. The difficulty noted by Wilbraham arises if and only if the probability of at least one of the single (or simple) events in a problem is unknown. in the case of the Challenge Problem the event E had unknown probability and this is the reason for the difficulty. It seems quite clear that this kind of difficulty would rarely arise in real-world problems so one can argue that this potential error is of little practical consequence. Nonetheless, it is possible to develop a simple procedure to handle this kind of situation. I believe that this procedure is entirely in accord with the structure of Boole's thinking and I have no doubt that he would have elaborated his theory in this general direction if he had recognized the merit of Wilbraham's stricture.

In the Challenge Problem we will find that there is an error that no one seems to have noticed. This error cannot be summarized briefly so discussion of it will be postponed until Chapter 4. However, it can be identified as resulting from Boole's treatment of conditioned variables. This error would be expected to be of infrequent occurrence but it is likely to occur more often than the difficulty noted by Wilbraham. in most cases this error can be avoided by careful statement of the assumptions of the problem.

For purposes of this paragraph let us assume that the difficulty discovered by Wilbraham is, in fact, an error. Then, it is ironic that the famous Challenge Problem incorporates both of the only two errors that have been discovered in Boole's system. This is, however, exactly the case. It is therefore important to emphasize that Boole's own method in many instances applies to important real-life examples where neither of the two errors arise. in fact, within my own experience it is rare to find either of the errors in actual applications of Boole's method. Clearly, however, it is desirable to call attention to the possibility of these two errors and to suggest how they can be eliminated. This will be done in Chapter IV of this book. (I should remind the reader that I call these "errors" because that is the common opinion. In my opinion, the first of these - the one emphasized by Wilbraham and Cayley - is not an error at all.)

In Chapter V we will first present a general statement about the kinds of data encountered in problems of advertising media selection. Then we can see that the simplified procedure that we will develop is sufficient as it stands for important and typical applications in such areas as advertising allocation based on available media data. It also has important potential applications in the analysis of contingency tables. Some examples of these uses of Boole's original method will be given in the final Chapter V. In addition, in Chapter V we will present some of the alternative methods that are available for obtaining exact probabilities in problems which do not have sufficient equations to permit the determination of unique values for all of the probabilities involved.

CHAPTER II

BOOLE'S LOGIC AND HIS PROBABILITY THEORY

Section 8. Guide to this Chapter

Most of those persons who were contemporaneous with Boole and familiar with his thinking took it to be a well-known fact that Boole constructed his logic primarily to provide a sound basis for his probability theory. This is sufficiently close to the reality to make it very difficult to understand Boole's probability theory without having a reasonably good foundation in his logical theory. This being the case, in this book it will be found that Appendix A, Boolean Algebra: A Brief Summary, provides a rapid overview of Boolean algebra which ought to be sufficient to enable the reader to follow the logical steps of Boole's method in probability theory. However, Appendix A will certainly not provide the background needed in order to follow the arguments, pro and con, about the validity of Boole's method.

Sections 9 and 10 of this Chapter are intended to provide the necessary level of Boole's logic for this purpose. Section 9, Boole's Logic, is a considerably longer presentation of Boole's logic than is Appendix A: specifically, Section 9 has 13 pages and Appendix A has only three pages. Further, the presentation of Section 9 attempts to highlight those logical theorems and propositions which turn out to be of specific importance in applications to Boole's probability theory. Section 10, Boole's Logical Analysis, Some Examples and Comments, gives some examples of Boole's logical analysis plus some discussion of the difficulties that may be encountered in undertaking such logical analyses. It is probably a fair statement that it will be difficult for the reader to follow the discussion of Boole's probability theory without having achieved some understanding of these two Sections.

Section 11, Probability: Boole's Theory and the Orthodox Theory Contrasted, already presents a first look at what may well be called the heart of the issue. As we shall see, Boole claimed that the orthodox probability theory could not answer many questions which Boole considered to be important ones and that his probability theory could answer these questions. Boole's opponents and critics denied Boole's contentions in this regard, maintaining that if the orthodox theory could not answer a question then it simply could not be answered. They argued, further, that Boole was surreptitiously introducing additional information which was not generally valid into his analyses and that this misled him into thinking that his theory could answer questions that the orthodox theory could not answer. The swirl of this argument about the differences between Boole's probability and the orthodox theory continued on many levels. In fact, the well-known criticism of Boole's theory by Wilbraham, sharpened considerably by Cayley, attacked Boole by arguing that his theory gave false answers to problems for which the orthodox theory gave correct answers. This Wilbraham-Cayley criticism was the one which led many persons to conclude that Boole's theory must be wrong. As we shall see, the criticism is itself wrong. In fact, it is not even coherent.

Section 12, Boole's Assumptions, briefly discusses the three assumptions that Boole made in developing his theory of probability. These are, first, that if the probabilities of two events are given and nothing is known about their dependence then we assume they are independent. Second, there is no basis for treating "simple" and "compound" events differently since the difference usually has nothing to do with the nature of the events but, rather, is primarily a matter of how we speak about them. Third, if events are not free to occur in every possible combination then Boole calls them "conditioned". This is the Boolean idea that Keynes misinterpreted so badly.

Section 13, Boole's Probability Theory, is an attempt to summarize all of the main ideas that Boole developed for his probability theory. We will be illustrating and discussing these ideas in most of the remainder of this book. It is worth mentioning here that Section 14, Boole's Method: A Verbal Transcription, is a less mathematical presentation of the same material and some readers may prefer to read Section 14 first. In fact, the subsequent four Sections are each the development of a part of Boole's procedure. Section 15, The Determination of Boole's V, is a discussion of a key quantity in Boole's theory; Section 16, The Argument from V to the Equations, continues the analysis; and Section 17, The Solution of the Equations, introduces an important simplification in finally achieving Boole's solution. Finally, Section 18, Boole's Procedure, Simplified and Summarized, gives an overall view of his whole theory of probabilities which makes relatively easy that which is usually quite difficult in Boole's analysis of probabilities.

In our statement of Boole's method we have introduced two major simplifications, one mentioned in the preceding paragraph. Keynes emphasized the great difficulty of Boole's method and even Boole himself referred to the calculations as "repulsive". The two simplifications here mentioned make relatively easy calculations that even Boole found very difficult. Section 19, An Example Comparing Boole's Method with the Proposed Emendations, undertakes to take a specific example and to show in detail how much easier the analysis is with the proposed emendations.

Finally, this Chapter ends with two Sections applying our simplifications to a number of Boole's own examples. Simpler illustrations are given in Section 20, Some of Boole's Examples, Based on Two Events, and somewhat more typical ones are given in Section 21, Some of Boole's Examples, Based on Three Events. Of course, throughout the remainder of this book there will be a variety of other examples, including further ones used by Boole.

Section 9. Boole's Logic

We know today that Boole's logical structure can be developed mathematically so that it is completely rigorous and, in fact, Professor Hailperin presents such a rigorous development in the first Part of his book. However, it certainly was not always so. As we pointed out in Section 3, algebraic logic never reached the textbook stage and the result was that there were serious doubts about the status of Boole's logical system. Most of the contributors to algebraic logic had challenged Boole's theory in one way or another. This is why the logic that developed from Boole's beginning is often called the Jevons-Peano-Schroder-Boole logic. This logic, as it is usually presented, resulted from the modifications introduced by other logicians. It was, until the appearance of Professor Hailperin's book, a vexed question whether, or to what degree, Boole's own logical structure was sound. Thanks to Hailperin's book, we now know that Boole's methods led to correct conclusions even if they were sometimes more complicated than necessary.

This is fortunate in terms of our purposes since we want to follow in Boole's footsteps, meaning that we want Boole's own logic as the foundation for his probability theory. We will summarize Boole's argument, as given in LT (An Investigation of the Laws of Thought). With his Proposition I (p. 27) he tells us that "All the operations of Language, as an instrument of reasoning, may be conducted by a system of signs composed of the following elements, viz.: 1st. Literal symbols, as x, y, etc., representing things as subjects of our conceptions. 2nd. Signs of operations, as +, -, *, standing for those operations of the mind by which the conceptions of things are combined or resolved so as to form new conceptions involving the same elements. 3rd. The sign of identity, =." Literal symbols are "appellative" or "descriptive" signs and express either the name of a thing or some quality belonging to it. Thus, let x be the class of things called "books". Further, let y be the class of things called "good" and let it be understood that a combination such as xy will represent the class of things to which x and y are simultaneously applicable. Then, xy will stand for "good books". Suppose, in addition, that z stands for "owned by the Detroit public library". Then xyz will represent "good books owned by the Detroit public library". Further, the order in which the signs occur does not affect the resulting class: xyz, zyx, yxz, and all other orders of the three signs signify the same class.

Boole then considers the laws to which the symbols such as x, y, etc. are subject. We have already noted that the order in which the signs occur does not affect the resulting class. This is usually summarized by stating that these literal symbols obey the commutative law, xy = yx. (The commutative law holds also of the symbols of ordinary algebra.) We can, therefore, use these symbols in place of the substantives, adjectives, and descriptive phrases, subject only to what Boole calls the interpretative law, that any expression in which several of these symbols are written together shall represent all the objects or individuals to which their several meanings are together applicable. (LT, pp. 29-30). If the two symbols in xy have exactly the same meaning - y having exactly the same meaning as x - then we would have xy = xx = x, or using the ordinary notation of algebra, X^2 = x and Boole emphasizes that this is a second general law of those symbols by which names, qualities, or descriptions are expressed. This simple relationship is one of the "signature" equations of Boolean algebra. After some discussion of the signs for mental operations, Boole will use this equation to develop his "law of duality".

Boole next (LT, p. 32f) considers signs of those mental operations whereby we collect parts into a whole or separate a whole into parts. The words we use in so doing are "and", "or": examples are "books and magazines", "hard-bound or paperback", "men and women". Let x = men and y = women and let + stand for "and" and "or". Then we have x + y = y + x, which is equally true if x and y represented numbers and + is the sign of arithmetical addition. Following Boole, let z stand for the adjective "European" and since it is the same thing in effect to say "European men and women" as to say "European men and European women", we have z(x + y) = zx + zy. Again, this holds true equally if x and y stand for numbers in ordinary algebra.

As Boole writes, "We cannot conceive it possible to collect parts into a whole, and not conceive it also possible to separate a part from a whole." in language we express this operation by the word "except", as in "all countries except absolute monarchies". It is ordinarily implied that the things excepted form a part of the things from which they are excepted. Since we have used + for aggregation it is natural, and convenient, to use - for "except". Thus, if x is taken to represent "countries" and y similarly represents "absolute monarchies" then "all countries except absolute monarchies" will be represented by x - xy. The meaning is indifferent to the order in which we write a series of terms, some of which are affected by -. Thus, x - y = -y + x. Continuing our present example, suppose that z represents "countries in the southern hemisphere". Then, "all countries in the southern hemisphere except absolute monarchies" would be represented by z(x - xy) = zx - zxy. These last equations illustrate the fact that the literal symbols x, y, etc. are distributive in their operation.

Boole next turns attention to the signs by which relation is expressed and by means of which we form propositions. He argues that every verb can be resolved into "is" or "are" and one of the literal signs and he expresses "is" or "are" by the sign =. Let x stand for "appropriate household pets", y is "dogs", and z is "cats" and "appropriate household pets are dogs and cats" will be represented by

x = y + z. If this is true then it will follow that appropriate household pets, except for dogs, are cats, or

$x - y = z$, This must be a deduction from the previous equation and we see that it is achieved by subtracting y from both sides of the equation. This illustrates the general axioms that: 1. If equal things are added to equal things, the wholes are equal; and 2. If equal things are taken from equal things, the remainders are equal. Further, if we have the equation $x = y$ then whatever z may represent we will also have $zx = zy$. This is formally the same as the algebraic law that if both members of an equation are multiplied by the same quantity, the products are equal. Similarly it may be shown that if the corresponding members of two equations are multiplied together, the resulting equation is true.

However, the analogy with algebra does not always hold. Suppose we have the equation

$zx = zy$. One's first thought is that in algebra we would divide through by z and obtain $x = y$. But a little closer thinking reveals that we can only do this if $z \neq 0$. Now, in Boolean algebra it is very often the case that z might = 0 so we cannot divide through by z. We can, nonetheless, solve the cited equation but a different route must be followed and will be presented below.

Boole wants next to determine the logical value and significance of the symbols 0 and 1. As used in algebra, the symbol 0 satisfies the formal law $0*y = 0$ (or simply $0y = 0$). The same law will be satisfied in logic if 0 is taken to be the symbol for Nothing, the Null class, and this is the assignment that Boole makes. Similarly, in algebra the symbol 1 satisfies $1*y = y$ (or simply $1y = y$) and Boole assigns the symbol 1 in his logic to the class that contains everything, the Universe. These assignments are Boole's Proposition II (LT, p. 47).

Granted that 1 is the symbol for the Universe, the class that contains all classes, then Boole's Proposition III follows immediately (LT, p. 48). Suppose x represents any class of objects. Then $1 - x$ will represent the contrary, or supplementary, class of objects, i.e. the class including all objects that are not included in the class x. Thus, if x represents "men" then the class 1, the Universe, can be conceptualized as being composed of all men and all not-men. The class "not-men" will be represented by $1 - x$.

We are now ready for Boole's Proposition IV (LT, p. 49). This is a result that plays an important role in Boole's logical theory and, hence, also in his probability theory. We start with the equation deduced above: $X^2 = x$. Subtract X_2 from both sides and we have $x - x^2 = 0$, which can be written as $x(1 - x) = 0$. The formal product of the expressions of two classes represents that class of individuals which is common to them both. The left-hand side of the last equation, then, is the expression for the class of individuals that are at the same time in the class x and in the class $(1 - x)$. But, as the right-hand side of the equation tells us, there are no individuals that are both x and not-x at the same time. This is the principle of contradiction, which affirms that it is impossible for any being to possess a quality and, at the same time, not to possess it. This is the principle which Aristotle maintained was the fundamental axiom of all philosophy and Boole provides a footnote in the original Greek and offers his own translation as well. The law shown by the equation $x(1 - x) = 0$ is often called by Boole the "law of duality". To verify it for a specific x, all that is necessary is to check whether $X^2 = x$.

Boole offers some further comments on the symbol + that is used for "and" and "or" and the symbol - that is used for "except". He writes that "and" is usually employed when the collection to be described forms the subject, "or" when it forms the predicate, of a proposition. Thus, "The scholar and the man of the world desire happiness" is an illustration of the former and "Things possessing utility are either productive of pleasure or preventive of pain" exemplifies the latter. He advises strongly that whenever an expression involving these conjunctions occurs in a proposition it is very important to know whether the groups or classes separated in thought by them are intended to be quite distinct from each other and mutually exclusive or not. Does the expression "scholars and men of the world" include or exclude those who are both? Does "either productive of pleasure or preventive of pain" include or exclude things which possess both of these qualities? Boole states that he thinks the formula "All x's are either y's or z's", when rigorously interpreted means "All x's are either y's, but not z's", or "z's but not y's". But, that stated, he says that ordinary ways of speaking seem to favor an opposite interpretation. By this he means that he suspects that the expression "Either y's or z's" would generally be understood to include things that are y's and z's at the same time, together with things that are y's but not z's and things which are z's but not y's. But the vitally important thing is that we must resolve any disjunctive expression that comes before us into elements that are really separated in thought, and then connect their respective expressions by the symbol +. From this it follows that, according to the meaning implied, the expression "Things which are either x's or y's" will have two different symbolic equivalents. If we mean "Things which are x's, but not y's, or y's but not x's" the expression will be $x(1 - y) + y(1 - x)$. If, however, we mean" Things which are either x's or, if not x's, then y's" the expression will be $xy + x(1 - y) + y(1 - x)$ which on multiplication out simplifies to $x + y(1 - x)$. Notice that the two expressions given do satisfy the law of duality. Thus, $\{x(1 - y) + y(1 - x)\}^2 = x(1 - y) + y(1 - x)$ and $\{x + y(1 - x)\}^2 = x + y (1 - x)$. It will be seen shortly that this is but a manifestation of a general law of expressions representing classes or collections of things.

Boole gives a Rule of Expression (LT, p. 57) in which he amplifies the examples of the preceding paragraph to three classes. Thus, "Either x's or y's or z's" would be expressed by $x(1 - y)(1 - z) + y(1 - x)(1 - z) + z(1 - x)(1 - y)$ when the classes denoted by x, y, and z, are designed to be mutually exclusive, and by $x + y(1 - x) + z(1 - y)(1 - x)$ when they are not designed to be exclusive.

On the rule of expression is founded the converse rule of interpretation, meaning how one is to understand the result of logical manipulations that are intended to achieve some purpose. Boole offers an example: let x = hard, y = elastic, and z = metals. Then:

"Non-elastic metals" will be expressed by $x(1 - y)$;
"Elastic substances with non-elastic metals" will be expressed by $y + z(1 - y)$;
"Hard substances, except metals by $x - z$;
"Metallic substances, except those which are neither hard nor elastic" by $z - z(1 - x)(1 - y)$ or else by $z\{1 - (1 - x)(1 - y)\}$;
"Hard substances, except those which are metallic and non-elastic, and those which are elastic and non-metallic" by $x - \{xz(1 - y) + xy(1 - z)\}$ or else by $x - xz(1 - y) - xy(1 - z)$.

These examples provide the background for Boole's Proposition III (LT, p. 58), which shows how to deduce from an examination of their possible varieties a general method for the expression of Primary Propositions, which Boole defines as propositions expressing a relation among things and to be contrasted with Secondary Propositions, which express a relation among propositions. Boole develops a rule for one kind of primary proposition: When the subject of the proposition is universal and the predicate of the proposition is also universal then form the separate expressions for them and connect them by the sign =. He suggests that this case is likely to arise in the expression of the definitions of science. As an example, he takes Mr. Senior's definition of wealth: "Wealth consists of things transferable, limited in supply, and either productive of pleasure or preventive of pain."[1] Boole remarks that wealth is really defined by its possession of three properties or qualities, not by its composition out of three classes or collections of objects. The conjunction "and", then, is superfluous. Let w = wealth,

t = things transferable, s = limited in supply, p = productive of pleasure, r = preventive of pain. Thus, the class of things which the above expression, taken alone, would define, would consist of all things productive of pleasure, together with all things productive of pleasure, but preventive of pain, and its symbolical expression would be p + (1 - p)r. We need the symbols s and t to limit its application to things "transferable" and "limited in supply", thus giving the symbolical equivalent for the original definition: w = st{p + (1 - p)r} (LT, p. 60). If the expression, "Either productive of pleasure or preventive of pain" were intended to point out those things which are productive of pleasure, without being preventive of pain, p(1 - r), or preventive of pain without being productive of pleasure, r(1 - p), the expression would be w = st{p(1 - r) + r(1 - p)}. Note that each of the two expressions for w observes the law of duality, $X_2 = x$.

Let us consider next the case where the subject of the proposition is universal while the predicate is particular. An example is the proposition, "All men are mortal", It is clear that the thrust of the proposition is "all men", a universal, are "some mortal beings", particular in the sense that it certainly is not intended to be a universal, "all mortal beings". We need an expression for "some mortal beings". We introduce v, a class that is indefinite in all regards except that some of its members are mortal beings. Let x be "mortal beings" and vx will represent "some mortal beings". Let y represent "men" and we have the equation y = vx. The Rule is (LT, p. 61): Express as before the subject and the predicate, attach to the latter the indefinite symbol v, and equate the expressions. Boole states that it is obvious that v is also a literal symbol, like x, y, etc., and that it meets the law of duality, $V^2 = v$. Remembering Boole's "exceptional mathematical intuition" we need not be dismayed if this is not obvious to us. We can accept the truth of the statement even without the obviousness.

Boole gives another example which is of future interest. He wants to symbolize the proposition, "The planets are either primary or secondary", so he lets x = planets (the subject), y = primary bodies, and z = secondary bodies and he recognizes that "or" separates absolutely the class of "primary" from that of "secondary" bodies. This means merely that yz = 0. Following the present rule, Boole finds for the equation of the proposition x = v{y(1 - x) + x(1 - y)} and he comments that the literal translation of the premises into the form x = v(y + z) is exactly equivalent but inferior because the first of these two equations meets the law of duality while the second does not. This is worthy of note because it typically means that subsequent steps in any analysis will be more onerous when the law of duality does not hold.

The next kind of proposition that Boole considers is universal negative propositions of which an example is "No men are perfect beings." Here we are certainly not speaking of a class termed "no men" and asserting that all the members of this class are "perfect beings". What is being asserted is that: "All men (subject) are (copula) not perfect (predicate)". So, if y represents "men" and x represents "perfect beings" we would represent this as y = v(1 - x) and similarly in any other case. Boole gives a rule

(LT, p. 63): To express any proposition of the form "No x's are y's", convert it into the form "All x's are not-y's" and then proceed as in the preceding case.

The final kind of proposition he considers is one in which the subject of the proposition is particular. An example is "Some men are not wise". This proposition does not mean that it is false that "Some men are wise." It does mean to say that "some men" lack wisdom. So the actual meaning of the proposition is "Some men are not-wise". Put y for "men" and x for "wise" and let v be a symbol of a class indefinite in all respects but that it contains some individuals of the class to whose expression it is prefixed and we have the representation of the proposition as vy = v(1 - x). Boole summarizes all of this in a General Rule for the Symbolical Expression of Primary Propositions (LT, pp. 63-4). He gives an example in illustration of the General Rule, the proposition "No men are heroes but those who unite self-denial to courage." Let x = "men", y = "heroes", z = "those who practice self-denial", w = "those who possess courage" and he paraphrases the proposition as "men who do not possess courage and practice self-denial are not heroes" with expression $x(1 - zw) = v(1 - y)$.

In Chapter V (LT, pp. 66-79) Boole begins developing the key ideas he uses for his logical theory and which, of course, are of critical importance for his probability theory. He begins with a question that was of major importance in the reception of his logical theory. Recall Lotze's very hostile opinion, summarized in Chapter I, about Boole's system of logic. His opinion was that only the premisses and conclusion of a logical argument by Boole were logical facts - everything between was meaningless. Certainly, Lotze was very extreme but many of the logicians who studied Boole's system were, to one degree or another, concerned about the same question. This question arose because Boole stoutly defended the position that, to quote Boole, ". . . it is an unquestionable fact that the validity of a conclusion arrived at by any symbolical process of reasoning, does not depend upon our ability to interpret the formal results which have presented themselves in the different stages of the investigation."

(LT, p. 68). In defending his position Boole often cited the use of the uninterpretable symbol $\sqrt{-1}$ in the intermediate processes of trigonometry. His critics argued the contrary position, that if any step of the argument produced a result that was uninterpretable then the conclusion could not be accepted. Boole offers three requirements in order to have valid reasoning by the aid of symbols (LT, p. 68) and these, of course, are in accord with his theory about interpretability. Boole tries to blunt the criticism by emphasizing that any system of propositions may be expressed by equations involving symbols x, y, z, etc., which, whenever interpretation is possible, are subject to laws identical in form with the laws of a system of quantitative symbols which can only assume the values 0 and 1. And he claims that this is exactly his procedure. He deduces the logical equations for a given problem, then lays aside the logical interpretation of the symbols in the equations, converts them into quantitative symbols taking only the values 0 and 1, performs on them as such all the necessary processes of solution, and finally restores to them their logical interpretation.

All of this would be of no help unless the final equations were in a form which would render their interpretation possible when the symbols are given their logical significance. Boole states, however, that there exists a general method of reducing equations to such a form and the rest of Chapter V is devoted to this method. He begins with the elementary observation that any algebraic expression involving a symbol x is termed a function of x and may be represented under the general form f(x). Similarly, if an expression involves two symbols, x and y, the general form would be f(x,y), and so forth for any number of symbols. Thus, the form f(x) could represent any of the following functions: x, $1 - x$, $(1 + x)/(1 - x)$, etc.; and the form f(x,y) could represent any of : $x + y$, $x - 2y$, $(x + y)/(x - 2y)$, etc. Following the same notational system, if we put $x = 1$ the result will be expressed by f(1) and if we put $x = 0$ the result will be expressed by f(0). Thus, if f(x) represents the function $(a + x)/(a + 2x)$ then f(1) will represent $(a + 1)/(a + 2)$, and f(0) will represent a/a.

We get a very important definition (LT, p. 72) of "developing" a function. Here is the definition: any function in which x is a logical symbol, or a symbol of quantity that can only take the values 0 and 1, is said to be developed (or expanded) when it is reduced to the form $ax + b(1 - x)$, where "a" and "b" are so determined as to make the result equivalent to the function from which it was derived. This leads to Proposition I: To develop any function f(x) in which x is a logical symbol. Assume that $f(x) = ax + b(1 - x)$. Making $x = 1$ we have $f(1) = a$. Similarly, making $x = 0$ we have $f(0) = b$. This shows that $f(x) = f(1)x + f(0)(1 - x)$ is the development sought. The right-hand side of the equation adequately represents the function f(x), whatever the form of that function may be, for if x is regarded as a quantitative symbol it can only assume the values 0 and 1 and for each of these the development assumes the same values as f(x). As an illustration, suppose we want to develop the function (1 + x)/(1 + 2x). We find f(1) = 2/3 and f(0) = 1/1 = 1. Hence the development is (1 + x)/(1 + 2x) = (2/3)x + 1 - x.

Proposition II (LT, p. 73) tells how to expand or develop a function involving any number of logical symbols. Consider some f(x,y). First, consider this to be a function of x alone and expand it by Proposition I: f(x,y) = f(1,y)x + f(0,y)(1 - x). Now take the coefficient f(1,y) and treat it as a function of y and get f(1,y) = f(1,1)y + f(1,0)(1 - y). Similarly, we find f(0,y) = f(0,1)y + f(0,0)(1 - y). Finally, substitute these last two expressions in the first equation and we have the desired development:

$$f(x,y) = f(1,1)xy + f(1,0)x(1 - y) + f(0,1)(1 - x)y + f(0,0)(1 - x)(1 - y) .$$

Thus, if f(x,y) = (1 - x)/(1 - y) we find f(1,1) = 0/0, f(1,0) = 0/1 = 0, f(0,1) = 1/0, and f(0,0) = 1, giving the development of the specified function as

$$(0/0)xy + 0x(1 - y) + (1/0)(1 - x)y + (1 - x)(1 - y) .$$

Boole will shortly show us that the forms 0/0 and 1/0 provide, in expressions such as the above, a very important logical interpretation. Proceeding similarly we can write the expansion for a function involving three symbols:

$$f(x,y,z) = f(1,1,1)xyz + f(1,1,0)xy(1 - z) + f(1,0,1)x(1 - y)z + f(1,0,0)x(1 - y)(1 - z) + \\ + f(0,1,1)(1 - x)yz + f(0,1,0)(1 - x)y(1 - z) + f(0,0,1)(1 - x)(1 - y)z + f(0,0,0)(1 - x)(1 - y)(1 - z) .$$

Each of the forms of expansion consists of certain forms into which the symbols x, y, z, etc. enter, multiplied by coefficients into which the symbols do not enter. The terms which contain symbols are called the constituents of the development. It is obvious that they are in form independent of the form of the function to be expanded.

The general rule of development will therefore consist of two parts, the first of which will relate to the formation of the constituents of the expansion, the second to the determination of their respective coefficients. Boole gives this general rule (LT, pp. 75-6) but I find his explanation unsatisfactory. Instead, to find the constituents for an expansion of a function involving n logical symbols, begin by recognizing that there will be 2^n constituents - thus 2 when there is one symbol, 4 when there are 2 symbols, 8 when there are 3 symbols, and so forth. The reason for this provides the simplest way to write out all of the constituents. Suppose there are 4 symbols: w, x, y, and z. If w were the only symbol then there would be two constituents, w and (1 - w). When we want the constituents for w and x we are going to add x and (1 - x) to each of the constituents with w alone, thus obtaining four: wx, w(1 - x), (1 - w)x, and (1 - w)(1 - x). When we want the constituents for w, x, and y we must add y and (1 - y) to each of the four constituents involving only w and x: wxy, wx(1 - y), w(1 - x)y, w(1 - x)(1 - y), (1 - w)xy, (1 - w)x(1 - y), (1 - w)(1 - x)y, and (1 - w)(1 - x)(1 - y). Finally, when we want the constituents for all four symbols we must add z and (1 - z) to each of the 8 constituents involving w, x, and y, thus beginning wxyz, wxy(1 - z), wx(1 - y)z, wx(1- y)(1 - z), and on to (1 - w)(1 - x)(1 - y)z, (1 - w)(1 - x)(1 - y)(1 - z).
Now, to find the coefficients for any constituent, proceed as follows. If the constituent contains w as a factor then change w into 1 in the original function. If the constituent contains (1 - w) as a factor then change w into 0 in the original function. For a given constituent make all of these changes, for each factor in the constituent, and the final calculated value of the function with these changes made is the coefficient of that constituent.

Boole comments (LT, p. 78) that sometimes functions do not necessarily become interpretable upon development, but equations are always reducible by this process to interpretable forms. It is worth an example to show what Boole means by "interpretable". As an illustration of this he considers the expression $x - y$. Suppose we develop $f(x,y) = x - y$. We find $f(1,1) = 0$, $f(1,0) = 1$, $f(0,1) = -1$, and

$f(0,0) = 0$. We have, then, the development $x(1 - y) - y(1 - x)$, since the other two coefficients are 0, and this is not interpretable. Why? Because we cannot take, in thought, from the class of things which are x's and not y's, the class of things which are y's and not x's, because the latter class is not contained in the former. Therefore, this development is not interpretable. But suppose we have the equation, $x - y = 0$. Then we would have upon development $x(1 - y) - y(1 - x) = 0$ and this is resolvable immediately into the two equations: $x(1 - y) = 0$ and $y(1 - x) = 0$, which tell us that there are no things which are x's and not-y's and no things which are y's and not-x's. Hence, $x - y$ does $= 0$.

Boole presents his Proposition III, which establishes some important properties of constituents

(LT, pp. 78-9). For simplicity, let t stand for any constituent. Then t satisfies the law of duality,

$t(1 - t) = 0$ or, alternatively, $t^2 = t$. Further, let t_1 and t_2 be any two constituents of a development. Then, the product $t_1t_2 = 0$. In words, the product of any two constituents of a development equals 0. Finally, the sum of all the constituents in any development equals 1. These results lead to what Boole calls Proposition IV (LT, p. 79). Let V = the sum of any series of constituents, where the separate coefficients are each equal to 1. Then V satisfies the law of duality, $V(1 - V) = 0$. All of these results are important in Boole's theory of probability.

Chapter VI (LT, pp. 80-98) is a very important Chapter on the general interpretation of logical equations. A complete development has two kinds of elements, the constituents, and their coefficients. In Chapter VI Boole first considers the interpretation of constituents and then investigates how that interpretation is modified by the coefficients of the constituents. Boole begins with his Proposition I

(LT, p. 81-2): The constituents of the development of any function of the logical symbols x, y, etc. are interpretable and represent the several exclusive divisions of the universe of discourse, formed by the affirmation and denial in every possible way of the qualities denoted by x, y, etc. Thus, consider just two symbols, x and y. The constituents are xy, $x(1 - y)$, $(1 - x)y$, and $(1 - x)(1 - y)$. These represent all the four classes of objects which can be described by affirmation and denial of the properties expressed by x and y. The classes are distinct from one another and they make up the whole universe. Upon the laws of constituents and the mode of their interpretation, as indicated in this paragraph, are founded the analysis and the interpretation of logical equations.

Boole proposes to investigate the forms of possible solution which present themselves in the conclusion of a train of reasoning. There are three such forms. The first form arises when any logical equation $V = 0$ is developed and the result is to be interpreted. Boole's Rule of interpretation for this form (LT; pp. 83-4) is: Develop the function V, and equate to 0 every constituent whose coefficient does not vanish. The interpretation of these results collectively will constitute the interpretation of the given equation. Boole gives an interesting example, based on the definition of "clean beasts" laid down in Jewish law, viz., "Clean beasts are those which both divide the hoof and chew the cud." He lets x = clean beasts, y = beasts dividing the hoof, and z = beasts chewing the cud. Then the given proposition is represented by $x = yz$, or $x - yz = 0$. Then, letting $f(x,y,z)$ represent $x - yz$ we find $f(1,1,1) = 0$, $f(1,1,0) = 1$, $f(1,0,1) = 1$, $f(1,0,0) = 1$, $f(0,1,1) = -1$, $f(0,1,0) = 0$, $f(0,0,1) = 0$ and $f(0,0,0) = 0$. So, there are four constituents which have 0 coefficient and, hence, four constituents that have non-vanishing coefficients.

According to the rule we must equate each of these last four constituents to 0. This gives $xy(1 - z) = 0$, $x(1 - y)z = 0$, $x(1 - y)(1 - z) = 0$, and $(1 - x)yz = 0$. These equations deny existence to the corresponding classes of objects: respectively, 1. Beasts which are clean, divide the hoof, but do not chew the cud; 2. Beasts which are clean, chew the cud, but do not divide the hoof; 3. Beasts which are clean, and neither divide the hoof nor chew the cud; and 4. Beasts which divide the hoof and chew the cud, but which are not clean. All of these denials are, in fact, implicit in the original proposition. Every primary proposition can thus be resolved into a series of denials of the existence of certain defined classes of things, and may, from that system of denials, be itself reconstructed.

The second form arises when any logical equation $V = 1$ is developed and the result is to be interpreted. Boole, in fact, uses the same example as for the first form and reasons in terms of the previous solution. Thus, in the preceding paragraph it was deduced from the logical equation that four constituents were denied existence. This means that the remaining four constituents represent classes that make up the universe. Therefore, $xyz + (1 - x)y(1-z) + (1 - x)(1 - y)z + (1 - x)(1 - y)(1 - z) = 1$ and this is equivalent to the affirmation that all existing things belong to one or another of the four following classes: 1. Clean beasts both dividing the hoof and chewing the cud; 2. Unclean beasts, dividing the hoof, but not chewing the cud; 3. Unclean beasts chewing the cud, but not dividing the hoof; and 4. Things which are neither clean beasts, nor chewers of the cud, nor dividers of the hoof. We can make a more general statement about the interpretation of this kind of logical equation as follows. If we have a logical equation $V = 0$ wherein V satisfies the law of duality, $V(1 - V) = 0$, then $1 - V = 1$. Thus, our V from the preceding paragraph gives immediately $1 - x + yz = 1$. If we set $f(x,y,z) = 1 - x + yz$ and develop the left-hand side we will get coefficients that will equal 1 - (coefficient when $V = 0$). Therefore, we will find that each of the four classes listed above has a coefficient of 1 and should be entirely selected. Stated alternatively, if we have a logical equation, $V = 1$, then we have immediately that $1 - V = 0$ and we can use the first form.

The third form arises when we need to interpret the equation $V = w$, where V is a function of the logical symbols x, y, z, etc. Boole states that this is a particular example of a problem of the "utmost generality" in logic: Given any logical equation connecting the symbols x, y, z, w; required an interpretable expression for the relation of the class represented by w to the classes represented by the other symbols x, y, z, etc. (LT, p. 87). He tells us that the solution of this problem consists in all cases in determining, from the equation given, the expression of the above symbol w, in terms of the other symbols, and rendering the expression interpretable by development. The equation given is always of the first degree with respect to each of the other symbols involved. The required expression for w can, therefore, always be found. In fact, if we develop the given equation, whatever its form may be with respect to w, we obtain an equation of the form $Ew + E'(1 - w) = 0$ where E and E' are functions of the remaining symbols. Therefore, $w = E'/(E' - E)$, and developing the right-hand side it only remains to interpret the result in logic. Of course, if $1 - w$ is desired we can immediately use $1 - w = E/(E - E')$. Boole's method is given by his Proposition III (LT, p. 88): To determine the interpretation of any logical equation of the form $w = V$, in which w is a class symbol, and V a function of other class symbols quite unlimited in form. Let V be developed. Each coefficient will belong to one of four classes:

1. Let the coefficient be 1. The whole of the class which it represents is implied.

2. Let the coefficient be 0. No part of the class represented by the constituent to which it is prefixed must be taken.

3. Let the coefficient be of the form 0/0. This is an indefinite class symbol, meaning that all, some, or none of the class to whose expression it is affixed must be taken.

To conclude what must be done with the fourth class, Boole states a theorem (LT, pp. 90-91): If a function V, intended to represent any class or collection of objects, w, be expanded, and if the numerical coefficient, "a", of any constituent in its development, does not satisfy the law of duality, $a(1 - a) = 0$, then the constituent in question must be made equal to 0. Boole shows generally that any constituent whose coefficient is not subject to the same fundamental law as the symbols themselves must be separately equated to 0. This gives:

4. Let the coefficient, "a", not satisfy $a(1 - a) = 0$ (the most common form under which such coefficients occur is 1/0), then set the corresponding constituent = 0. The reality is that if the coefficient is different from any of the three possibilities above then it is usually this case.

It follows, then, that if the solution of a problem, obtained by development, be of the form

$$w = A + 0B + (0/0)C + (1/0)D$$

then that solution may be resolved into two equations: $w = A + vC$ and $D = 0$, where v is an indefinite class symbol. The interpretation of the first of these two equations shows what elements enter, or may enter, into the composition of w, the class of things whose definition is required. The interpretation of the second of these two equations shows what relations exist among the elements of the original problem, in perfect independence of w.

Returning to the example of "clean beasts", required a description of "unclean beasts". As before, x = "clean beasts", y = "beasts dividing the hoof", and z = "beasts chewing the cud". We have $x = yz$, whence $1 - x = 1 - yz$. Let $f(y,z) = 1 - yz$ and developing $f(y,z)$ we find $f(1,1) = 0$, $f(1,0) = 1$, $f(0,1) = 1$, and $f(0,0) = 1$. Therefore, $1 - x = y(1 - z) + (1 - y)z + (1 - y)(1 - z)$ which can be interpreted: Unclean beasts are all which divide the hoof without chewing the cud, all which chew the cud without dividing the hoof, and all which neither divide the hoof nor chew the cud.

Starting with the same definition, required a description of beasts which do not divide the hoof.

From the same definition we have $y = x/z$ and, hence, $1 - y = (z - x)/z$. Put $f(x,z) = (z - x)/z$ and develop $f(x,z)$. We find $f(1,1) = 0$, $f(1,0) = -1/0$, $f(0,1) = 1$, and $f(0,0) = 0/0$. Whence we must set $x(1 - z) = 0$ and we have $1 - y = 0\,xz + (1-x)z + v(1 - x)(1 - z)$ and $x(1 - z) = 0$. The first of these two equations gives beasts which do not divide the hoof consist of all unclean beasts which chew the cud and an indefinite remainder (none, some, or all) of unclean beasts which do not chew the cud. The second equation gives the proposition, there are no clean beasts which do not chew the cud. Independently of any relation to beasts dividing the hoof, there exists, in virtue of the premiss, a separate relation between clean beasts and beasts which chew the cud.

We are now ready for the last two major additions to Boole's logic, ones that are particularly important for his probability theory. These are mainly developed in Chapter VII, "On Elimination",

(LT, pp. 99-113) and Chapter VIII, "On the Reduction of Systems of Propositions", (LT, pp. 114-129). We will take them in order. Chapter VII deals with the elimination of some symbols from a logical equation. Boole points out that in algebra one is generally able to eliminate (n - 1) symbols from n equations. No such relationship exists between the number of equations given which represent propositions or premisses and the number of symbols which can be eliminated. From a single equation an indefinite number of symbols may be eliminated. When symbols need to be removed from logical equations we often refer to them as "middle terms" because the elimination of "middle terms" is the main step in finding the intended conclusion of a syllogistic argument. Here is an old example. We have two premisses: All men are mortal, and Athenians are men. What can be concluded? Let x represent "men", y represent "mortal", and A represent "Athenians". Then we have two equations: $x = vy$ and $A = v'x$, where v and v′ are each an indefinite class symbol but distinguished since they do not necessarily have the same value. Let us write the first equation as $x - vy = 0$. Let $f(v) = x - vy$. Then, to eliminate v from this equation all we need do is calculate f(1) and f(0) and multiply them together. Here we find $f(1) = x - y$ and $f(0) = x$; the product is $f(1)f(0) = x - yx = x(1 - y) = 0$ and it is clear that v has, indeed, disappeared from the equation. Similarly, we write the second equation as $A - v'x = 0$ and take the left-hand side to be f(v′); We find $f(1) = A - x$ and $f(0) = A$ so the product is $A(1 - x) = 0$ and we see that v′ has been eliminated. As we shall see shortly, we are entitled to add these two equations together, giving $x(1 - y) + A(1 - x) = 0$. Let the left-hand side be represented by f(x) and we want to eliminate x. We find $f(1) = 1 - y$ and $f(0) = A$, giving the product $A(1 - y) = 0$ and this is the desired conclusion. $(1 - y)$ is the class of immortal things and $A(1 - y)$ represents the Athenians who are members of this class of immortal things and we see that no Athenians are in this class. In other words, all Athenians are mortal.

Boole's Proposition I (LT, p. 101) covers exactly the argument that we have just illustrated: If $f(x) = 0$ is any logical equation involving the class symbol x, with or without other class symbols, then the equation $f(1)f(0) = 0$ will be true, independently of the interpretation of x; and it will be the complete result of the elimination of x from the equation. Similarly, the complete result of the elimination of any class symbols x, y, etc., from any equation of the form $V = 0$, will be obtained by completely expanding the left-hand side of that equation in constituents of the given symbols, and multiplying together all the coefficients of these constituents, and equating the product to 0. The second part of this Proposition covers situations in which one wishes to eliminate several symbols at once. Perhaps it is simpler to state the procedure differently. Suppose we have some equation, $V = 0$, and that V is a function of x, y, and several other logical variables. We are still entitled to consider the left-hand side to be represented by f(x,y). Then, using the notation as before, if we calculate f(1,1), f(1,0), f(0,1), and f(0,0) then $f(1,1)f(1,0)f(0,1)f(0,0) = 0$ will be the original equation with x and y eliminated from it.

To illustrate the general method of Chapter VII, Boole returns to his example based on Mr. Senior's definition of wealth, "Wealth consists of things transferable, limited in supply, and either productive of pleasure or preventive of pain." Boole uses w to stand for wealth, s to stand for things limited in supply, t to stand for things transferable, p to stand for things productive of pleasure, and r to stand for things preventive of pain. This gives the equation $w = st\{p + r(1 - p)\}$. As Boole states, we can eliminate any symbols from this equation that we do not desire to take into account and express the result by solution and development. First, we write the equation in the form $w - st\{p + r(1 - p)\} = 0$. Now suppose we want to find the expression for w if we eliminate r. As usual, writing the left-hand side of the equation as f(r) we find $f(1) = w - st$ and $f(0) = w - stp$. Multiplying these together we find $(w - st)(w - stp) = w - wstp - wst + stp = 0$. We see that r has been eliminated from the equation. We solve for w and find $w = stp/(st + stp - 1)$. By developing the right-hand side we find that the coefficient of the constituent stp is 1 and the coefficient of the constituent $st(1 - p)$ is 0/0. All of the other six constituents have coefficient 0. Therefore, our interpretation is that wealth consists of all things limited in supply, transferable, and productive of pleasure, and an indefinite remainder of things limited in supply, transferable, and not productive of pleasure.

We should note here that this procedure is of vital importance for Boole's theory of probability. In a typical problem there will be various events represented in our equation for which we do not know the probabilities. Boole's first step will be to eliminate the symbols representing those events from the equation. This permits, or enforces, that we will have the unknown probability, for which we wish to solve, as a function only of events for which we know the probabilities.

As another example, let us eliminate s and t from the logical equation for Senior's definition of wealth: $w = st\{p + r(1 - p)\}$. We want to eliminate s = things limited in supply and t = things transferable, and having done so, let us find the definition of $(1 - p)$ = things not productive of pleasure. Writing the equation in the form $w - st\{p + r(1 - p)\} = 0$ and representing the left-hand side as $f(s,t)$, we find $f(1,1) = w - p - r(1 - p)$, $f(1,0) = w$, $f(0,1) = w$, and $f(0,0) = w$. We must multiply these four terms: $f(1,1)f(1,0)f(0,1)f(0,0) = w - pw - rw + rpw = 0$ or $w(1 - p - r + rp) = 0$. We can easily solve this for p and we find $p = \{w(1 - r)\}/\{w(1 - r)\}$ whence $1 - p = 0/\{w(1 - r)\}$. Expanding the right-hand side, we find four constituents: $w(1 - r)$ with coefficient 0, and wr, $(1 - w)r$, and $(1 - w)(1 - r)$, each with coefficient 0/0. Therefore, $1 - p = \{0/0\}[wr + (1 - w)r + (1 - w)(1 - r)] = \{0/0\}[wr + (1 - w)]$, as Boole writes the answer [LT, p. 112].

It remains to summarize Boole's argument in Chapter VIII, "On the Reduction of Systems of Propositions" [LT, pp. 114-129]. The basic idea is easily stated. Boole wants to find a procedure so that a set (or system) of equations can be reduced to a single equation, thus permitting the application of the preceding results to that equation - in particular, all of the results of Chapter VII. Boole tells us that there will be shown in Chapter VIII two distinct modes of reducing systems of equations to equivalent single equations. He says that the first method has the advantage of being simple in theory but the disadvantage of rendering the subsequent processes of elimination and development "somewhat tedious". Nonetheless, he proceeds to present the first method in Proposition I [LT, p. 115-120]: Any system of logical equations may be reduced to an equivalent single equation, by multiplying each equation after the first by a distinct arbitrary constant, and adding all the results, including the first equation, together. By this he means that he has shown the following: Suppose the general system is $V_1 = 0$, $V_2 = 0$, $V_3 = 0$, etc.. Then Boole shows that the general system may be replaced by $V_1 + cV_2 + c'V_3 +$ etc. $= 0$, where c, c', etc. are arbitrary constants. The equation thus formed may be treated in all respects as the ordinary logical equations of the previous chapters. The arbitrary constants c, c', etc. are not logical symbols. They do not satisfy the law of duality.

The second method is the subject of Proposition III (LT, p. 121): If $V_1 = 0$, $V_2 = 0$, etc., the terms of which have been brought by transposition to the left-hand side, then the combined interpretation of the system will be involved in the simple equation $(V_1)^2 + (V_2)^2 +$ etc. $= 0$, formed by adding together the squares of the given equations. As a corollary, any equation, $V = 0$, of which the left-hand side already satisfies the law of duality, $V^2 = V$ or $V(1 - V) = 0$, does not need, since it would be unaffected by, the process of squaring.

Proposition V (LT, p. 122) is Boole's attempt to present a "practical rule" or method for the reduction of systems of equations expressing propositions in logic. For this purpose he returns to the division of propositions he already developed for his general method of expressing primary propositions and which we summarized on pp. 4-5 above. We will list them again here:

1. Propositions of which the subject is universal and the predicate particular. The symbolical type is $X = vY$ where X and Y satisfy the law of duality and v is the indefinite symbol. Eliminating v, we have $X(1 - Y) = 0$, which also satisfies the law of duality. No further reduction by the process of squaring is needed.

2, Propositions of which both terms are universal. The symbolical type is X = Y where X and Y separately satisfy the law of duality. Writing the equation as X - Y = 0 and squaring we get X(1 - Y) + Y(1 - X) = 0 and the left-hand side obviously satisfies the law of duality.

3. Propositions of which both terms are particular. The symbolic form of such propositions is vX = vY but v is not quite arbitrary and therefore should not be eliminated. For, as Boole tells us, v is representative of some, which, though it may include in its meaning all, does not include none. So, we must write the equation as vX - vY = 0 and square, obtaining vX(1 - Y) + vY(1 - X) = 0, which does satisfy the law of duality.

Boole then summarizes this in his Rule (LT, p. 124) and he adds that any equation which is given in the form X = 0 will not need transformation, and any equation which presents itself in the form X = 1 may be replaced by 1 - X = 0.

Given a single equation which is equivalent to some system of equations, we have seen how to solve for any specific logical symbol in terms of the other symbols but this is not sufficiently general to meet Boole's statement of his goal in developing his logical system. He makes a statement about the general problem (LT, pp. 140-143): Given any equation connecting the symbols x, y, . . . , w, z, . .; Required to determine the logical expression of any class expressed in any way by the symbols x, y, . . In terms of the remaining symbols, w, z, . . . Boole's derivation of his result is somewhat complicated so I will merely call attention to the pages involved and already cited and summarize by giving Boole's Rule for accomplishing this purpose (LT, p. 142): Expand the given equation with reference to the symbols x, y. Then form the equation Et + E′(1 - t) = 0, in which E is the product of the coefficients of all those constituents in the development in terms of x and y, whose coefficients in the expression of the given class are 1, and E′ the product of the coefficients of those constituents of the development whose coefficients in the expression of the given class are 0. The value of t deduced from the cited equation by solution and interpretation will be the expression required.

This rule seems to me to be less perspicuous than most of Boole's rules and procedures. Perhaps we can make it clearer. We can illustrate the procedure by assuming that there are a total of four logical symbols and we want to determine the logical expression in terms of w and z of some class defined in terms of x and y. This will require that we make two developments in terms of x and y. One will be of the class, which is defined in terms of x and y, and the other will be the development of the left-hand side of the given equation (V) in terms of x and y. The constituents for both of these developments will be xy, x(1 - y), (1 - x)y, and (1 - x)(1 - y). The key equation we need to solve the problem is Eu + E′(1 - u) = 0. where, for convenience we have put u = given class. Boole proves that the coefficients of the constituents in the development of the class will each be either 0 or 1. He then shows that E will be the product of the coefficients of the constituents in the development of V, which constituents have coefficient 1 in the development of the class. Similarly, E′ will be the product of the coefficients of the constituents of V, which constituents have coefficient 0 in the development of the class. This gives the needed equation and all that must be done further is to solve for u and obtain the development and interpret it in logical terms.

Boole again gives an example based on Mr. Senior's definition of wealth: w = st{p + r(1 - p)}, wherein w = wealth, s = things limited in supply, t = things transferable, p = things productive of pleasure, and r = things preventive of pain. Required is an expression for things transferable, but not productive of pleasure, t(1 - p), in terms of w, s, and r. We develop t(1 - p) in terms of t and p: t(1 - p) = 0 tp + t(1 - p) + 0 (1 - t)p + 0 (1 - t)(1 - p). We obtain our given equation by squaring the two sides of w - stp

- str(1 - p) = 0, giving w + stp + str(1 - p) - 2wstp - 2wstr(1 - p) = 0. We now develop the left-hand side of the last equation in terms of t and p, obtaining after a little rearranging:

{w(1 - s) + s(1 - w)}tp + {w(1 - sr) + sr(1 - w)}t(1 - p) + w(1 - t)p + w(1 - t)(1 - p) = 0 .

Only the coefficient of t(1 - p) is 1 in the development of the class so E = w(1 - sr) + sr(1 - w), the coefficient of t(1 - p) in the development of the equation. The other three coefficients are each 0 in the development of the class so E′ = product of the corresponding coefficients in the development of the equation, or E′ = {(w(1 - s) + s(1 - w)}{w}{w} = w(1 - s). For convenience, let u = t(1 - p). The desired equation, then, is

{w(1 - sr) + sr(1 - w)}u + w(1 - s)(1 - u) = 0
whence
u = [w(1 - s)]/[2wrs - rs - sw] .

We need to develop u in terms of r, s, and w. We find four constituents with coefficient 0/0: wsr, (1 - w)s(1 - r), (1 - w)(1 - s)r, and (1 - w)(1 - s)(1 - r); two constituents with coefficient 0: ws(1 - r) and (1 - w)sr; and two constituents with coefficient 1/0: w(1 - s)r and w(1 - s)(1 - r). Hence, things transferable and not productive of pleasure are either wealth, limited in supply and preventive from pain; or things which are not wealth, but limited in supply and not preventive of pain; or things which are not wealth, and are unlimited in supply. Note that this interpretation correctly amalgamates the last two of the constituents with 0/0 as coefficient. This is because (1 - w)(1 - s){r + (1 - r)} = (1 - w)(1 - s).

Section 10. Boole's Logical Analysis, Some Examples and Comments

Our goals in this Section are to give some examples of Boole's logical analysis and to discover what major difficulty in that procedure is the biggest contributor to the "notorious difficulty" of this kind of analysis. Remember the previously quoted comment of Keynes: "... the difficulty of his method is so great that I do not know of any one but himself who has ever attempted to use it." Is that true? If so, what is the specific cause? As we will see, in the case of Boole's probability analysis we can tremendously simplify the difficulties. As far as I can determine, no comparable improvement is possible for his logical analysis but there are many logical problems where the difficulty, to be specified, does not arise.

Boole claimed that y = v(1 - xz) and yx = v(1 - z) are equivalent (LT, pp. 105-6). In order to prove this we need only eliminate v from each equation, separately. Writing the first equation as y - v(1 - xz) = 0 we treat the left-hand side as f(v) and we want f(1), which equals y - 1 + xz, and f(0), which equals y. We do this because f(1)f(0) = 0 will be the original equation with v eliminated. This is also a reminder that v is subject to the law of duality. We find f(1)f(0) = y - y + xyz = xyz. Proceeding similarly with the second equation, we have yx - v(1 - z) = 0 and we take the left-hand side of this equation as f(v), whence f(1) = yx - 1 + z and f(0) = yx. Therefore, f(1)f(0) = yx - yx + xyz = xyz. We see that, as Boole claimed, the two expressions are equivalent.

The most typical of the kinds of problems that are appropriate for a Boolean formulation are ones which have several (two or more) statements of input information and, thus, have two or more initial equations. Here is a fairly straightforward problem of this kind:

If X that is not A is the same as B, and A that is not X is the same as C, what is X in terms of A, B, and C?[2]

Here it is quite easy to write the original equations, as given in the above information. X that is not A is XA′ and "the same as B" gives the equation, XA′ = B. Similarly, from the second such statement we have X′A = C. The immediate goal in all problems with two or more initial equations is to amalgamate the equations into one equation and then develop the expression found for the variable of interest. First, we must write the equations transposed so that all the logical variables are on the left-hand side. Boole tells us exactly how this should be done for three major classes of equations, including the present type. These equations are of Boole's generic form X = Y and where X and Y each satisfy the duality requirement, which is evidently true here, as can be seen by straightforward multiplication. Thus, XA′*XA′ = XA′ and B*B = B, and similarly for the other equation. In this case, Boole tells us that the resulting equation will be X(1 - Y) + Y(1 - X) or, for our specific values, XA′(1 - B) + B(1 - XA′) = 0 and, for the second equation, X′A(1 - C) + C(1 - X′A) = 0.

Since this amalgamation step is of key importance, Boole gives no less than three ways to accomplish it. One of the most useful methods is this: given $V_1 = 0$ and $V_2 = 0$, convert to the equation, $(V_1)^2 + (V_2)^2 = 0$. Further, if V_1 and V_2 meet the requirement of duality we do not have to square them. Instead, we can simply write $V_1 + V_2 = 0$. Note that we can directly test for duality by multiplication. Thus, $\{XA'(1 - B) + B(1 - XA')\}^2 = XA'(1 - B) + B(1 - XA')$ and similarly for the other equation. Therefore, our single equation will be, writing, for convenience, (1 - X) for X′

$$XA'(1 - B) + B(1 - XA') + (1 - X)A(1 - C) + C(1 - (1 - X)A') = 0 .$$

We are interested in X so we want to solve this equation (but no division!) for X:

$X[A'B' - A'B - AC' + AC] + B + AC' + C - AC = 0$, where we have written B′ for 1 - B. Hence,

$$X = \frac{AC - B - AC' - C}{AC + A'B' - A'B - AC'}$$

We now have to develop this expression for X. There are eight constituents and I find it quite easy to arrange the necessary information in tabular form:

Constituent	A	B	C	Num	Den	Coefficient
ABC	1	1	1	-1	1	
ABC′	1	1	0	-2	-1	
AB′C	1	0	1	0	1	0
AB′C′	1	0	0	-1	-1	1
A′BC	0	1	1	-2	-1	
A′BC′	0	1	0	-1	-1	1
A′B′C	0	0	1	-1	1	
A′B′C′ 0	0	0	0	1	0	

("Num" is numerator of X, "Den" is denominator of X.) We are only interested in coefficients of 1 but I have also shown two coefficients of 0. A coefficient of 1 means that that constituent is to be included completely in the equation for X, which, hence, is

$$X = AB'C' + A'BC' .$$

In words, X equals all A's that are not B's or C's plus all B's that are not A's or C's. Can we check this answer? in many examples we are able to check the presumed answer but this is increasingly difficult to do as the original problem becomes more complex. Our problem is a simple one so we would expect to be able to check the answer. As a preliminary we can notice three true statements that hold because of the original equations stating the given information:

from $XA' = B$, multiplying both sides by A gives $AB = 0$,

from $X'A = C$, multiplying the right side by B and the left side by XA' gives $BC = 0$,

from $X'A = C$, multiplying both sides by A gives $X'A^2 = X'A = AC$ whence $AC = C$
whence $C(1 - A) = CA' = 0$.

According to our information, XA' should $= B$ but $XA' = A'(AB'C' + A'BC') = A'BC' =$
$(1 - A)B(1 - C') = (1 - A)(B - BC) = (1 - A)B = B - AB = B$, as it should.
Also according to our information, $X'A = C$ but $X'A = (1 - AB'C' - A'BC')A = A - AB'C' =$
$A(1 - (1 - B)(1 - C)) = A(B + C) = AC$.
We already know that $AC = A$ but if we proceed as usual we have $X'A = AC$. To solve for C we must develop $X'A/A$ (remember, no division!) And we find that the constituent $X'A$ is the only one with coefficient 1 so by development $C = X'A$, which was to be proved. Therefore, the expression for X that we found does, indeed, meet the requirements of our initial information.

Boole specifically states that the method of amalgamating equations used above (going from $V_1 = 0$, $V_2 = 0$, $V_3 = 0$, etc. to $(V_1)^2 + (V_2)^2 + (V_3)^2 +$ etc.) is a general method for any system of equations of the indicated type (LT, Proposition III, pp. 121-2). He gives another general method; multiply each equation after the first by a distinct arbitrary constant and add all the results, including the first equation, together. If we apply this method to our example, letting k be the constant multiplier for the second equation, we find that the resulting amalgamated equation is

$$XA'B' + B - XA'B + k(1 - X)AC' + kC(1 - A + XA) = 0\,.$$

This gives the expression for X as

$$X = \frac{B + k(AC' + C - AC)}{A'B + kAC' - kAC - A'B'}$$

If we develop X we find, again, that the only constituents with coefficient 1 are the same two as we found above and again we conclude that $X = AB'C' + A'BC'$.

The third method Boole gives is not as well presented as one might prefer. Further, it is fairly involved. We will not illustrate it here. Boole presents this method in (LT, pp. 152-3).

One of the main reasons why so many of Boole's examples are very involved is that he was anxious to demonstrate the power of his logical algebra. In order to do this convincingly he often stated problems as examples that were complicated, involved five or more logical variables, and included conditions that would be unlikely to arise in most real-life problems. A good illustration of the difficulties one encounters in working on Boole's examples is his Example 5 in Chapter IX

(LT, pp. 146-9). He states the problem thus: A class of natural productions has led to the following general results:

1. In any of these productions if the properties A and C are missing, the property E is found, together with one of the properties B and D, but not with both.

2. Whenever the properties A and D are found while E is missing, the properties B and C will either both be found, or both be missing.

3. Whenever the property A is found in conjunction with either B or E, or both of them, there either the property C or the property D will be found, but not both of them. Conversely, wherever the property C or D is found singly, there the property A will be found in conjunction with either B or E, or both of them.

Before we translate these statements into suitable symbols and equations we can notice that five different properties are involved. This immediately means that the algebra is likely to be troublesome. Boole asks several different questions about this example but we will be satisfied to take only his first question, which is to determine what the occurrence of A means in terms of the other variables.

However, for all of his questions, including the one we have cited, he adds a typical Boolean complication. This is that he wants to say nothing at all about E in any conclusions. In other words, E is to be eliminated from further consideration. Boole often adds such a condition and it is somewhat surprising because this kind of restriction surely does not arise very often in real-world problems. One can think of some instances where it might arise but they are quite unusual. There appear to be two major reasons why Boole so often introduces this kind of condition. First, his logical algebra makes it quite straightforward to eliminate variables so he might be anxious to demonstrate this power. But second, this kind of elimination of variables is a crucial step in his theory of probabilities. We shall see that his whole treatment of probability problems directly involves this kind of elimination.

We will need an indefinite class symbol for the first two equations and we shall use q for this purpose, as Boole does in this example. (He ordinarily uses v but this conflicts with his notation in the present example.) The first statement can be rendered as $A'C' = qE(BD' + B'D)$. In order to eliminate q we need to write the equation as $A'C' - qE(BD' + B'D) = 0$. Letting the left-hand side be f(q), we need $f(1) = A'C' - EBD' - EB'D$ and $f(0) = A'C'$. The product, f(1)f(0), is the desired equation, with q eliminated. Here we find 1) $A'C' - A'BC'D'E - A'B'C'DE = 0$. For the second statement we start with $ADE' = q(BC + B'C')$ and take the same steps as for (1). So doing, we find that the second equation is 2) $ADE' - ABCDE' - AB'C'DE' = 0$. The point of the third statement is to state an equality between two defined classes. Therefore, we do not need an indefinite class symbol. Instead, we can write the equation, $AB + AB'E = C'D + CD'$. This is one of the forms for which Boole has given the way to convert it so that it meets the requirements of the duality principle. So doing, we have the third equation, 3) $(AB + ABE')(CD + C'D') + (C'D + CD')(1 - AB - ABE') = 0$. (in writing this we have used the fact that $(1 - C'D - CD') = CD + C'D'$.)

In any such analysis with more than one equation, the first step must be to amalgamate the several equations into one equation. All three of these equations meet the duality principle - you can verify this for the first and second equations by direct multiplication - so we are entitled simply to add the three equations together and set the sum = 0. This, then, is the equation from which the analysis begins:

$$A'C' - A'BC'D'E - A'B'C'DE + ADE' - ABCDE' - AB'C'DE' + ABCD + ABC'D' + AB'CDE + AB'C'D'E + C'D + CD' - ABC'D - ABCD' - AB'C'DE - AB'CD'E = 0 .$$

What we now need to do is clear. First, we must eliminate E from this equation by the usual method. Let the whole left-hand side of the equation be f(E). Find f(1) and f(0) and the product, f(1)f(0), will be the desired equation without E. Solve the resulting equation for A in terms of B, C, and D and A will be the ratio of two logical expressions involving only B, C, and D. Then develop the ratio in terms of the eight constituents of the three logical variables. Interpreting this development will show what is known about B, C, and D when it is known that A occurs.

We will give some more thought to the question of the difficulties of the various steps shortly. However, I can simply state here that the work involved in getting the above single equation was not particularly onerous. Roughly, an hour was probably all it took, including checking the original Boole references to make sure that the translation of the three initial equations was done correctly. (Note that this statement is somewhat misleading if it is generalized but we will clarify the issue shortly.)

Working from the above equation, however, and trying to obtain that which is summarized immediately below the equation, is far more time-consuming than one would ever imagine it to be. It is a remarkably error-prone process. I spent all of my spare time for a week and I never got the right answer. And notice that I had a big advantage in that I accepted that Boole's answer was undoubtedly correct so all I had to do was to compare my answer with his in order to conclude that I was wrong again and so must start over. Usually, in such analyses, there is no obvious way to find the answer by some alternative method of analysis and thus verify that your Boolean argument is correct. Boole, himself, could not escape this difficulty. The last sentence of Chapter IX is referring to this Example 5 when he writes,

"I have not attempted to verify these conclusions." Nor have I and, furthermore, I don't know how to verify the conclusions. It is usually possible to verify conclusions when there are a lesser number of logical variables but it is always unlikely to be accomplished when there are five or more variables.

So, what am I going to write about this Example since I "never" got the right answer? When my level of frustration reached a sufficient height I returned to Boole to see how he did it. It immediately was apparent that he avoided the whole analysis that I was trying to accomplish. Chapter IX is entitled "Methods of Abbreviation" - meaning, ways to avoid the long algebraic manipulations that I was trying to accomplish. He uses Example 5 to illustrate the method of analysis suggested by his Proposition III,

(LT, p. 133). I had thought that Proposition III would save, at best, a relatively small proportion of time so I had skipped any study of it. It is not appropriate to spend much space in expositing it here since we have no need of it in Boole's probability theory. However, it is worth summarizing its application to Example 5.

Suppose t is any variable that is retained in the final result of the elimination of any other symbols from an equation. Thus, for Example 5, t could be anyone of A, B, C, and D since all of these are in the final result of eliminating E. Then the final result of such elimination may be expressed in the form Ft + F′(1 - t) = 0, in which F is formed by making t = 1 in the equation and then eliminating E and F′ is formed by making t = 0 in the equation and then eliminating E. It sounds as if the net process is worse since we will have to go through the elimination process twice but let us see. It is more convenient to write our single equation in the form before multiplying the terms out, which is easier to work with:

A′C′ - A′BC′D′E - A′B′C′DE +ADE′ - ABCDE′ - AB′C′DE′ + (AB + AB′E)(CD + C′D′) +
(C′D + CD′)(1 - AB - AB′E) = 0.

Now, we will use A as our "t". To find F, put A = 1 and get

DE′ - BCDE′ - B′C′DE′ + (B + B′E)(CD + C′D′) + (C′D + CD′)(1 - B - B′E) .

We must eliminate E from this by the usual procedure: treat the whole expression as f(E) and find f(1) = CD + C′D′ and f(0) = CD + C′D′ + B′C′D + B′CD′. The product, f(1)f(0) = CD + BC′D′, is F. To find F′ put A = 0 in the single equation above and get C′ - BC′D′E - B′C′DE + C′D + CD′. F′ is obtained by eliminating E from this equation. We find

$$F' = C' - BC'D' - BC'D + 2C'D - B'C'D + CD' .$$

In accordance with Proposition III, we now know that $FA + F'(1 - A) = 0$. Solving for A, we have immediately that $A = F'/(F' - F)$. This, of course, if written out is a function of B, C, and D and we want to find the development of A in terms of the constituents of the three logical variables. Obtaining this expression for A was enormously easier than the straightforward approach with which I exhausted a week of spare time. This was completely contrary to my original (uninformed) opinion.

Remember that the coefficient of a constituent is obtained by substituting the values of the three logical variables in the expressions for F and F′. Thus, BCD has a coefficient given by substituting

$B = 1$, $C = 1$, and $D = 1$, giving $F = 1$ and $F' = 0$ so the coefficient is $0/1 = 0$ and none of BCD is included in A. BCD′ requires substituting $B = 1$, $C = 1$, and $D = 0$, and has coefficient $1/1 = 1$ so BCD′ is included when A occurs. We find that BCD′, BC′D, B′CD′, B′C′D, and B′C′D′ each has a coefficient of 1 so we conclude, as Boole does, that

$$A = BCD' + B'CD' + BC'D + B'C'D + B'C'D' = CD' + C'D + B'C'D' .$$

Verbalizing this relationship, we can say that if A occurs then either C or D occurs, but not both, or else all three of B, C, and D will not occur. Since we have an equality it follows that the converse also is true. This means that if one of C and D occurs, but not both, or if none of B, C, and D occurs then A will also occur. This is exactly the answer that Boole obtained so we have every reason to be satisfied.

Since Boole presents quite a diversity of kinds of examples it is dubious that any one example can be called a "typical" one. This granted, we may nonetheless usefully compare this example with many of the others with particular regard to three dimensions. First, the algebra involved in the use of Boole's logical analysis is often at least as annoying as this Example 5. To eliminate a variable becomes more and more complicated as the number of logical variables increases. But even with few variables, if the initial equation is a lengthy one then considerable algebraic manipulations will be required - and very often of a type that is particularly error-prone. Second, we finally concluded in our Example that one of Boole's methods of abbreviation might be usefully applied. This is rarely an obvious characteristic of a given logical problem. Remember Boole's reputation for having a fine "mathematical intuition": he apparently was able to recognize possible applications of one or another of his Propositions more readily than most of us can. It is easy to get the impression that Boole developed some one of the methods of abbreviation for a particular example and then presented it as if it might frequently be applied. It is quite rarely the case that in attempting a given logical problem the insight suddenly arises that some method of abbreviation might usefully be tried.

The third dimension is one for which our Example is atypical. It is a general fact that one of the most difficult steps in solving logical problems is to correctly formulate the initial equations that summarize the input information. This did not arise with our Example 5 because we had the input equations as Boole wrote them. Obviously, this can only happen if the analyst is analyzing one of Boole's own examples. In all other cases - the common situation - the analyst hopes, but is rarely certain, that he or she has the initial equations correct. Notice how this question compounds with the other dimensions in a way that often leads to enormous algebraic difficulties. For each way of writing the initial equations there will be the same heavy algebra and some kind of "answer" will be achieved. But is it correct? If we don't have a known answer then how do we ever know if the analysis is correct?

Here is an example of a problem that "ought" to be made to order for a Boolean logical analysis:

A set of balls has each ball spotted with one or more of the colors red, green, and blue. The balls are also numbered. All of the balls spotted with red are also spotted with blue. All the odd-numbered blue balls and all the even-numbered balls which are not both red and green are on the table. Describe the balls that are not on the table.2

The major difficulty that I discovered in this problem was the question of how to write the logical equations that are equivalent to the information given. It seemed to me that there ought to be a "neat" way to write the equations so that Boole's method, so to speak, would do all the "work". If there is such a way to write the equations then I never found it. This was not for want of trying. I count 27 sheets of laborious algebra before I found an answer that seems to be correct. Further, there is nothing the least bit "neat" in my solution. I used an obvious notation: B for blue, G for green, R for red, E for even numbers, and T for on the table. Then I translated the input information in this way: BG'R' = vT, EB'GR'= vT, E'B'GR' = vT', and so forth, where v is the indefinite class symbol. There are eight such equations. I eliminated the v from each of them, added them all together in one equation = 0, and solved this equation for (1 - T) = T'. Then I developed the expression for T' in terms of the 16 constituents for E, B, G, and R. The result is that T' = 1 - T = E'BGR + E'BGR' + E'B'GR' + E'BG'R. In words, the balls not on the table are the odd-numbered balls spotted blue and red, and either green or not, and the odd-numbered balls spotted green and not-red, and either blue or not.

This was a lot of work, no matter how one adds it up. Furthermore, I am still not convinced that the answer is correct. How can one be convinced? It seems to me that there are three possible routes: 1. The problem is from some collection of such problems and the answer is given; 2. It is possible to work out the solution by some method other than Boole's; 3. The analyst has complete faith in his or her ability with Boole's system. In actual practice, the second possibility is the only one that can be relied on but it is often impossible to accomplish. And so we reach the conclusion that the heaviest algebraic load in a Boolean logical analysis is very often the first step: writing down the equations that translate the input information into usable form.

Section 11. Probability: Boole's Theory and the Orthodox Theory Contrasted

It is impossible to over-emphasize the fact that Boole maintained that the orthodox theory of probability did not provide means for finding solutions to a variety of problems that Boole considered to be very important ones. He made the point as clearly and as emphatically as he could but almost all his commentators either ignore, or fail to comprehend, Boole's statements in this regard. Hailperin and Keynes, to take two examples, often use the orthodox theory to calculate probabilities in one or another example used by Boole. They then announce that Boole did not get the "right" answer and perhaps they add that anyhow he was close. In fact, if Boole had found the same answers as those found by the orthodox theory then he would have totally failed in his purpose. Boole was attempting to solve that which he called "the general problem of probabilities" and that problem certainly cannot be solved by the orthodox theory. Let us begin by giving Boole's statements on this difference.

Boole's statement of the general problem of probabilities, the problem that he proposed to solve, was previously quoted but it is worth quoting it again (LT, p. 15)

> "Given the probabilities of any events, simple or compound, conditioned or unconditioned: required the probability of any other event equally arbitrary in expression and conception."

Elsewhere he states it somewhat differently (LT, p. 246)

"Given the probabilities of any events, of whatever kind, to find the probability of some other event connected with them."

The second statement, obviously equivalent to the first, is made after Boole notes that, strictly speaking, any combination of events is also an event. This truth is more important for Boole's theory than it might appear and it will be more lengthily discussed in the next Section.

Boole discusses the fact that it is equally possible to consider that probabilities refer, not to events, but to the propositions which assert that those events have occurred (LT, p. 247). With this interpretation, numerical probabilities have reference to the truth of the propositions rather than to the occurrence of the events. This distinction is not particularly important for our purposes but it does represent the format in which Boole chose to summarize the orthodox theory. After giving a summary, taken mostly from Laplace, of the "principles that have been applied to the solution of questions of probability", seven in number, Boole offers the following summary of the kinds of questions that can be solved by orthodox probability theory (LT, p. 250):

DATA : 1. The probabilities of n conditional propositions:

If the cause A_1 exist, the event E will follow;
If the cause A_2 exist, the event E will follow;

. . . .
If the cause An exist, the event E will follow.

2. The condition that the antecedents of those propositions are mutually conflicting.

REQUIREMENTS :

The probability of the truth of the proposition which declares the occurrence of the event E; also, when that proposition is known to be true, the probabilities of truth of the several propositions which affirm the respective occurrences of the causes A_1, A_2, ..., An.

Boole emphasizes that being able to solve this general problem is not at all equivalent to being able to solve the general problem of probability as he has formulated it. He specifically calls attention to the facts that the orthodox theory requires that the antecedents of the propositions should be mutually exclusive and that there is only one consequence, E, in the whole system. To his view, these are very restrictive assumptions.

To the original seven principles, Boole adds an eighth, due to W. F. Donkin, Savilian Professor of Astronomy at Oxford University (LT, p.251):

"If there be a number of mutually exclusive hypotheses, H_1, H_2, H_3, . . . of which the probabilities relative to a particular state of information are P_1, P_2, P_3, . . . and if new information be given which changes the probabilities of some of them, suppose of h_{m+1} and all that follow, without having otherwise any reference to the rest; then the probabilities of these latter have the same ratios to one another, after the new information, that they had before, that is,

$p_1' : p_2' : p_3' : . . . : p_m' = P_1 : P_2 : P_3 : . . . : p_m$

where the accented letters denote the values after the new information has been acquired.

Boole considers that this proposition is more fundamental than the others and he suggests that it might even be taken to be axiomatic. Now, in fact, this principle plays an important role in Boole's theory and seems to be involved in one of the two possible errors mentioned in Section 7. However, in the present context Boole simply wants to note that even adding this principle to the other seven would not in any way change the generality of the kind of problem that can be solved by orthodox probability theory. In order to accomplish this purpose other principles are needed.[3]

Section 12. Boole's Assumptions

Boole's system is based on three "principles", or assumptions, that he discusses in the beginning of Chapter XVII of An Investigation of the Laws of Thought. None of the assumptions is very far removed from generally accepted ways of thinking although the legitimacy of their application in particular circumstances may be challenged.

The first assumption is (LT, p. 256):

> "When the probabilities of events are given, but all information respecting their dependence withheld, the mind regards them as independent."

He states the same assumption again thus (LT, p. 256):

> "If my knowledge of two simple events is confined to this particular fact, viz., that the probability of the occurrence of one of them is p, and that of the other is q; then I regard the events as independent, and therefore affirm that the probability of their joint occurrence is pq."

Obviously, if the probability of the joint occurrence of the two events is known then this is part of the data of the problem. But if we know nothing about this joint probability then, Boole is claiming, we naturally assume that the two events are independent. The appropriateness of this assumption will be discussed shortly but it certainly does not seem much out of accord with common sense.

The second assumption is not much more startling than the first one. This assumption refers to the distinction we often make between simple and compound events (LT, p. 255):

> "I would, in the first place, remark that the distinction between simple and compound events is not one founded in the nature of the events themselves, but upon the mode or connexion in which they are presented to the mind."

This is a point to which Boole ascribes considerable importance and he returns to it quite often in his writings. Essentially, he says that that which we call a simple event is likely to be something that can be stated in a simple sentence. Therefore, he argues, the "simple" is mostly the result of an accident of language rather than some differentiation between the events themselves. He instances as an example the statement "to prosper", which might be regarded as expressing a "simple event" even though a large number of separate particulars are implied by the statement.

Boole's best example of this point occurs in the paper for which he was awarded the Keith Medal.[4] He states a problem:

> "The probability of the concurrence of rain and snow is p, of the concurrence of snow and wind q and of the concurrence of wind and rain r; required the probability of the concurrence of wind, rain, and snow."[5]

Now he supposes that in some dialect of English there is no separate word for rain or for snow or for wind but there is a word, sleet, for the concurrence of rain and snow; there is a word, drift, for the concurrence of snow and wind; and there is a word, storm, for the concurrence of wind and rain. Then, according to Boole, the problem would be stated:

> "The probability of sleet is p, of drift q, and storm r; required the probability of the concurrence of sleet, drift, and storm."[6]

But this does not express all of the information that the person using the dialect would possess because he would have to know that it was impossible that any two of the events sleet, drift, and storm could occur without the third also occurring. Therefore, finally, such a person would state the problem thus:

> "The probability of sleet is p, of drift q, of storm r, and these events are so connected, that no two of them can occur without the third occurring. What is the probability of their concurrence?"[7]

Boole claims (and refers to it as a "principle" in the reference quoted) that "the solution of the problem must be the same whichsoever of these forms of statement we adopt".[8] The argument that underlies Boole's position would undoubtedly proceed as follows. A correct solution of a problem in probability cannot depend on the language in which the problem is formulated. To argue otherwise would be to introduce nominalism into a realistic science. Based on this argument, Boole takes the stated "principle" as being generally accepted. Therefore, this "principle", which he accepts as being obviously true, is not the point of the example he has constructed.

His second assumption is not the "principle" of the preceding paragraph. Rather, it is another point illustrated by the same example. Returning to the difference between simple and compound events, most English-speaking persons would think of the original formulation of the problem (in terms of rain, wind, and snow) as involving simple events. The formulation in terms of sleet, drift, and storm would appear to be in terms of compound events. But, as Boole is emphasizing, the person who thinks in terms of the dialect would consider that sleet, drift, and storm were simple events. For that person, "snow" would be the concurrence of sleet and drift, so "snow" would seem to be a compound event. For English-speaking persons, "sleet" seems to be compound since it is the concurrence of snow and rain. But Boole argues, and this example is intended to illustrate, that the distinction results only from language and has nothing to do with the events themselves.

If one accepts Boole's argument, and it seems reasonable, then his assumption follows. This is that in a method for dealing with the general problem of probabilities there is no basis for treating "simple" and "compound" events differently. In particular, and the major implication for Boole's method, the first assumption applies equally well to all events. Therefore (LT, p. 256):

> "When the probabilities of events are given, but all information regarding their dependence withheld, the mind regards them as independent. And this mode of thought is equally correct whether the events, judged according to usual expression, are simple or compound, i.e., whether each of them is expressed by a single verb or by a combination of verbs."

Before proceeding to the third, and last, assumption, I will make a comment on the second assumption. This is that despite the cleverness of Boole's constructed example, it is certainly not clear that his argument is correct and, in fact, it appears quite dubious. It would not be appropriate here to become involved in complex questions about languages but Boole's example must be challenged. For English-speaking persons it is quite clear that rain, snow, and wind are simple events and sleet, drift, and storm are certainly compound events constructed from the three simple events. However, for the persons speaking Boole's hypothetical dialect it is not at all clear that sleet, drift, and storm would be simple events. If these persons are to be able to define the usage of these words they must be aware that the words are defined in terms of rain, snow, and wind and therefore that these are the simple events. For a person speaking the dialect, Boole suggests that "snow" would be the concurrence of sleet and drift and, hence, "snow" would appear to be a compound event. However, this is not correct because storm always occurs, given the concurrence of sleet and drift, and this means that rain, snow, and wind are each equally involved in the concurrence of sleet and drift (and, hence, storm). Boole's argument based on this example is curious because he applies his method to each of the two versions of the problem, as stated above, and obtains different answers. He then tries to show that one of the two is "right" but his argument here is not sound. We will discuss his analyses subsequently. As a preliminary summary, it is fair to say that this challenge to his second assumption does not significantly affect his method. However, we should note here that when we look at these two examples in Chapter IV we will discover a very important reason for concluding that the two analyses are not equivalent. In fact, it seems extremely unlikely that Boole could have overlooked the evidence that the two analyses summarized here are quite different in their conclusions. It is only fair to state in Boole's behalf that he actually refers to three different versions and it is perhaps most reasonable to assume that his comments, excerpted here, refer to the two versions of the dialect example and do not refer to the difference between the dialect examples on the one hand and the English-speaking example on the other.

The third assumption is also illustrated by the preceding example. It may be the case that, in addition to the probabilities of various events, there is definite information respecting the possible combinations of the events. Thus, when the previous problem was formulated in terms of sleet, drift, and storm then it was known that it was impossible for any two of these three events to occur without the third one occurring also. When this kind of situation obtains, Boole says that the events in question are "conditioned". Precisely, when events are not free to occur in every possible combination they are called conditioned. It is in this context that Boole makes the statement that was misunderstood by Keynes (see Section 4). The statement is: "Simple unconditioned events are by definition independent." (LT, p. 258) Boole is obviously referring to events for which only the probabilities of the events taken singly are known. In this case, by his first assumption one assumes that the events are independent.

Boole summarizes the effect of these assumptions thus (LT, pp. 256-7):

> "The events whose probabilities are given are to be regarded as independent of any connexion but such as is either expressed, or necessarily implied, in the data; and the mode in which our knowledge of that connexion is to be employed is independent of the nature of the source from which such knowledge has been derived."

This somewhat enigmatic statement deserves to be amplified by a paraphrase of it also given by Boole:

"The mode in which we avail ourselves of this information in the calculation of the probability of any conceivable issue of events depends not upon the nature of the events whose probabilities and whose limits of possible connexion are given. It is indifferent from what source, or by what methods, the knowledge of their probabilities and of their connecting relations has been derived. We must regard the events as independent of any connexion beside that of which we have information, deeming it of no consequence whether such information has been explicitly conveyed to us in the data, or thence deduced by logical inference." (LT, p. 256)

The point of the two preceding quotations is that it is of no consequence whether we are explicitly given information concerning, for example, some impossible combinations of events or whether we have deduced it by logical analysis from the given information. In either case the information will be used in exactly the same way. Of course, the main key to Boole's method in probability is the use of his logical system. This will be the subject of the next Section.

Before considering how Boole proposes to use his logical system in the resolution of the general probability problem as stated by him, it is worth discussing the reasonability of his assumptions. Suppose we are considering events A and B and know only that p(A) = p and p(B) = q. If we know nothing whatsoever about the conditional probabilities of A given B or of B given A then most persons would probably agree that our best available estimate of the probability of AB would be pq. We would, in short, do as Boole says we must: if we have no information to the contrary we assume the events are independent. There are two possible objections to this approach that, I think, really miss the point so they should be noted. First, it is clearly possible - and, indeed, might frequently be the case - that we have some knowledge that suggests that the two events are not independent but not enough knowledge to estimate the joint probability. Suppose, for example, that A = female and B = color-blind. We know that p = .5 and q = .05 but we do not know p(AB). It is well known that color-blindness is sex-linked so we have reason to be quite doubtful that we should estimate p(AB) = pq = .025. However, if we have no additional information - including, for example, whether females have more or less color-blindness than males - then we may have no alternative. On the other hand, if p(AB) enters into an analysis of great importance then we should invest some resources to find a more adequate estimate. In any case, this is surely not a defect in Boole's argument.

Second, suppose we are considering five different events and all that is known is the single probability of each of them occurring, regardless of the occurrence or not of any of the others. Then by Boole's argument we should estimate the probability that all five events occur simultaneously as the product of the five individual probabilities. I think that in fact this is correct but it needs to be emphasized that no rational person is going to have much confidence in the truth of such an estimate. However, this kind of situation is due entirely to our lack of information. The use of Boole's assumption - equivalent to the commonsense one - will produce estimated probabilities for each of the 32 possible combinations of occurrence of the five events and these will undoubtedly be the best such estimates, given the information available. We will legitimately have little confidence in them but this is an example of those problems that arise in the area Keynes called the weight of arguments.[9]

There are perhaps, some might say probably, many kinds of probability problems for which one would not want to make the above assumptions. I do not know of such problems but I make the statement in order to show that I am not attempting to defend the general applicability of Boole's methods to all probability problems. It is sufficient for my purposes that there are important problems for which his assumptions are completely apt. It is perhaps worth showing one such. This is the problem that arises in the analysis of media data, where A and B are two magazines and the probabilities are that a randomly selected household will have seen A, B, or both. For some actual data p = .161, q = .229, and

p(AB) = .075. Since .161*.229 = .037 it is clear that A and B are not independent. The actual probability of AB is about twice as much as it would be if A and B were independent. This is true for the corresponding data for almost any pair of magazines. In other words, it is unusual to find any pair of magazines that act independently. There are many possible reasons why this is true and they are not hard to find. We know that many households have adults that are functionally illiterate. Thus, the probability that a household reads any other magazine is increased when it is known that the household reads some given magazine. Again, a household that reads one magazine shows that it takes pleasure from reading, that it enjoys reading magazines, that it will spend money to get a magazine, and so forth. Each of these characteristics suggests that the probability that the same household reads another magazine is increased. These are a few of the many reasons why one would not expect the readership of a pair of magazines to be independent random events.

Now, suppose we are considering five such magazines and we have, as we usually do, data concerning each pair of the five (ten pairs). Each pair taken by itself will show dependence just as the pair above. However, the reasons remain similar for all the pairs. Perhaps, then, we might expect that the data by pairs exhausts the dependence and that combinations of three or more of the five magazines would be independent, granted the dependence by pairs. I am not aware of a good vocabulary to state my point clearly but I hope my general thrust is clear. Suppose combinations of three magazines show dependence additional to that shown by the three pairs within the combinations of three. Then, surely, the magazines in combinations of four or more should be independent granted the dependence by pairs and the additional dependence by threes. At some point, by threes or by fours or by fives, the dependence simply has to be exhausted and the magazines ought then to act as if they were independent (beyond the number that exhausts the dependence). To think otherwise is to assume, in effect, that there is unlimited complexity in these data. Such an assumption is anti-scientific and I do not know of any instance where it has been found to be true. In fact, it is almost by definition impossible.

The point of this example is that the poor verbalization given of what needs to be done in this case is exactly what Boole proposes to do. Suppose that the dependence among magazines is exhausted by taking the magazines in threes. It is possible to obtain data that give all of the single magazine probabilities, all of the probabilities by pairs, and all of the probabilities by threes. Using Boole's methods on such data is equivalent to assuming that all combinations of four or more magazines can be calculated from the given data under the assumption that for four or more magazines the data can be assumed to be independent.

This is Boole's assumption and this is exactly what one wants to do with these data. There is no other method in the theory of probability that accomplishes this purpose.

Section 13. Boole's Probability Theory

(Note that Section 14 is a less mathematical statement of the method that is being discussed in this Section. Some readers may prefer to read Section 14 before reading this Section.)

The idea of treating simple events as if they were independent when nothing is known to the contrary is surely not very profound. If such an obvious approach were the essence of Boole's method then, indeed, fools might be made professors of it. But, of course, this is not the case. Boole certainly intends to treat simple events as if they were independent when nothing is known to the contrary but he also intends to treat compound events in exactly the same way. This is in accord with his assumption that there is no inherent difference between simple and compound events. The essence of the problem, then, is: Exactly how is this to be done?

Boole's answer is that his algebra of logic is precisely the tool that is needed. It is almost always easy for intellectual successors to over-estimate the "obviousness" of some key step made by an originator. Recognizing that this is a risk, it may nonetheless be true that in the present instance it did not require a stroke of genius to imagine that logic might "naturally" be applied to probability. One could find a fair number of quotations by logicians other than Boole to the effect that, after all, logic, dealing with truth and falsity, was analogous to probability where the events were neither certain nor impossible. According to Keynes, Leibniz, in his article on the mode of electing the kings of Poland, conceived of probability as a branch of logic.[10] Since this article dates to 1669 it is evident that there is quite a respectable history behind this key step in Boole's procedure. This granted, it is important to remember that Boole was the first to actually attempt to do it.

It is easy to find simple illustrations of the relationships between Boole's algebra of logic and elementary probability theory. Consider the possible combinations of two simple events, x and y, with respective probabilities p and q. Boole offers the following tabulation (LT, p. 257):

EVENTS		PROBABILITY
xy	Both x and y	pq
x(1-y)	x and not-y	p(1-q)
(1-x)y	not-x and y	(1-p)q
(1-x)(1-y)	not-x and not-y	(1-p)(1-q)

We will often write the negation of x as x' but it needs to be remembered that Boole included both subtraction and division in his algebra and that he represented the negation of x by (1-x). In any case, it is clear that the correspondence between his algebra of logic and the usual calculations of joint probabilities is exact. The tabulation is of conjunctive combinations of the constituent events. Consider a disjunctive combination such as "the occurrence of either the event x without the event y or the event y without the event x". This would be represented in Boolean algebra as x(1-y) + (1-x)y. The probability is p(1-q) + (1-p)q. In other words, in these various instances the probability is the same function of p and q as the logical expression is of x and y.

Boole shows that this is a general law. Let there be any unconditioned event V, some function of events x, y, z, . . . with respective probabilities p, q, r, . . . Then the probability of V will be found by changing in V the symbols x, y, z, . . . Into p, q, r, . . . Following Boole, we will denote the result of this substitution as [V]. Under the same circumstances, suppose there is another event W. Suppose it is desired to determine the probability (conditional in the usual sense of the word) that if V occurs, W will also occur. Then Boole shows that the desired probability is [VW]/[V], where VW is the logical product of V and W. This is Boole's Proposition I (LT, p. 258). The requirement that V should be unconditioned means that there is no logical restriction on the joint occurrence of events x, y, z, . . . This is the first proposition and it is in accord with Boole's stated intention (LT, p. 258): "We shall proceed in the natural order of thought , from simple and unconditioned, to compound and conditioned events."

As an elementary illustration, consider events x, y, and z with probabilities p, q, and r, respectively. Suppose it is required to find the probability that if either x or y occurs singly then either y or z will occur singly. Then V = x(1-y) + (1-x)y and W = y(1-z) + (1-y)z. Using the ordinary rules of Boolean algebra (see Appendix I) we multiply these and find VW = x(1-y)z + (1-x)y(1-z). Hence, [VW] = p(1-q)r + (1-p)q(1-r) and [V] = p(1-q) + (1-p)q. The desired probability is the former divided by the latter. This example is given by Boole (LT, p. 260). The answer is obviously correct and some experimentation will soon convince anyone that Proposition I does work as Boole asserts.

Following Boole, suppose now that V is, as above, some function of the single events x, y, z, . . . but now V is going to be a condition that has been satisfied when the probabilities of x, y, z, . . . were determined to be p, q, r, . . . What, then, are the "absolute probabilities" of the events x, y, z, . . .; that is, the probabilities of their respective occurrence independently of the condition V? Boole's answer to this question is what he intends to be his Proposition II. His statement concerning both the underlying question and the proposed answer is not perspicuous and Wilbraham took specific exception to it. Perhaps for this reason, Boole added a Note after the Table of Contents. In the Note he explains that by the "absolute probabilities" of the events x, y, z, . . . he means "what the probabilities of those events ought to be, in order that, regarding them as independent, and their probabilities as our only data, the calculated probabilities of the same events under the condition V should be p, q, r, . . ." (LT, Note after Table of Contents)

So construed, let p′, q′, r′, . . . be the desired probabilities. Let [V]′ be the result of substituting p′, q′, r′, . . . for x, y, z, . . . In V. Designate those terms in V which have x as a factor by V(x) and similarly for y and z, giving V(y) and V(z). Then, letting [V(x)]′ designate the substitution of p′, q′, r′, and so forth into V(x), and similarly for V(y) and V(z) and by Proposition I we can write the equations:

$$[V(x)]'/[V]' = p; \quad [V(y)]'/[V]' = q; \quad [V(z)]'/[V]' = r; \quad \ldots.$$

Since there are as many equations as there are unknowns we can solve for the desired p′, q′, r′, and so forth. This simple procedure is what Boole intends as Proposition II although he does not state this clearly.

The kind of situation to which this argument refers is not one that is very likely to arise in real problems. The most obvious situation to which it would apply is the case where probabilities have been determined by a frequency count but that for some reason not all of the results have been included in the count. Even in this case the problem has to be specialized to the case where the joint events are not all recorded separately. Boole's example is one in which balls are being drawn from an urn with the balls being either marble or white and he specifies that no count is kept of those cases where the ball was neither white nor marble. We will comment on this example subsequently.

Here is a different example that is analogous to Boole's. Suppose a mail-order company has a file of all the customers to which it has sent an order in the past year. During the year the company has offered two different merchandise premiums to its customers and it has separate files showing the customers who obtained each of the two premiums. They now make a count as follows. Randomly selecting a name from the customer file they have clerks verify whether it is in either of the premium files. If the customer is not in either premium file he or she is not counted and no separate count is maintained of the number of customers in both premium files. Let x designate the event that a customer is in the first premium file and y that the customer is in the second premium file. Suppose p is the proportion of names counted for x and q is the proportion of names counted for y. Then Boole's Proposition II applies, with

$$V = xy + (1-x)y + x(1-y).$$

Then V(x) = xy + (1-x)y = x and, similarly, V(y) = y. Therefore, [V(x)]′ = p′ and [V(y)]′ = q′. Since [V]′ = p′q′ + (1-p′)q′ + p′(1-q′) we immediately have two equations:

$$\frac{p'}{p'q' + (1-p')q' + p'(1-q')} = p \text{ and } \frac{q'}{p'q' + (1-p')q' + p'(1-q')} = q$$

Solving these two equations gives the required answer:

$$p' = \frac{p + q - 1}{q} \quad \text{and} \quad q' = \frac{p + q - 1}{p}$$

Note that it is necessarily true that p + q ≥ 1. For the present problem this would be Boole's "condition of possible existence".

The reader may well be wondering why this amount of space has been devoted to such an unlikely kind of problem. The reason is, as mentioned above, that this example is a different version of Boole's example based on an urn with the balls drawn being either marble or white (LT, p. 262). This is the specific example to which Wilbraham took exception (in addition to Boole's solution of the Challenge Problem) and we will subsequently see that this example is closely related to a possible error in Boole's system, as mentioned in Section 7.

Boole's Proposition III states how, following his method, one can express the event whose probability is sought as a logical function of the events whose probabilities are known. This is perhaps the major conceptual step in his procedure. Suppose that x, y, z, . . . are simple events and that S, T, . . . are compound events that are known functions of x, y, z, . . . and whose probabilities are known. Let W, also some function of x, y, z, . . ., be the event whose probability is sought. Following Boole's argument that there is no real distinction between simple and compound events, we are going to replace S, T, . . . (which may be complicated expressions involving x, y, z, . . .) with single logical symbols s, t, . . ., w. Thus:

s = S; t = T; . . . . ; w = W .

These are a series of logical equations connecting the symbols s, t, . . . , w with the simple events x, y, z, . . . Now, by using the procedures of Boole's algebra of logic we can eliminate from the above system of equations any of x, y, z, . . . whose probabilities are not known and then determine w as a developed function of s, t, . . . and any of the simple events with known probabilities. It follows, then, that this expression for w will be the complete expression of the relation sought. It will present w as a function of simple events with known probabilities and of compound events with known probabilities. And, of course, by using s, t, . . . Boole has treated the compound events S, T, . . . as if they were simple events, in accordance with his basic assumptions.

The procedure is perfectly straightforward. However, it quickly becomes very onerous when the number of simple and compound events increases. Everyone, including Boole, agrees that his method, as he presented it, requires heavy algebraic manipulations. There are two main reasons for this and the present step is one of them. The next example will follow Boole's method as he presented it (and as he used it). I note that Appendix II gives another such example. In both examples, and generally in Boole's own examples the algebra becomes disconcertingly heavy. Therefore, I will emphasize here that we will shortly find ways to very much reduce the algebra required.

A typical example will show the procedure and suggest the difficulties. Suppose we are in advertising and are considering three magazines, A, B, and C. We know p(A) = p and p(B) = q but we do not know p(C). We also know p(AB) = r and p(BC) = s and we need to estimate p(ABC). (These are standard kinds of measurements in advertising: p(A) and p(B) are the reaches of the two magazines, meaning that p, for example, gives the proportion of some specified target audience that is reached by A; p(AB) and p(BC), called intersections, give the proportion of the target audience reached by both specified magazines; and p(ABC) is the proportion reached by all three specified magazines.) The original simple events are A, B, and C and we know the probabilities of two of these. We also know the probabilities of two compound events: U = AB and V = BC. Finally, we want to know the probability of W = ABC. Following Boole's procedure, we define three simple events by the equations:

$$u = AB \qquad v = BC \qquad \text{and} \qquad w = ABC\ .$$

We convert each of these to an equation equal to zero using the standard Boolean procedure. Thus, u = AB becomes uA′ + uB′ + ABu′ = 0, where A′ = (1 - A) and similarly for the other symbols. Adding the above equations to the two others (from v and from w) gives

$$uA' + uB' + ABu' + vB' + vC' + BCv' + wA' + wB' + wC' + ABCw' = 0$$

The next step is to eliminate C from this equation since we do not know the probability of C. In Appendix A it is shown that this is accomplished by the same procedure that is used to write the normal form of the equation (for C). The algebra of this step is quite a bit easier if we take out for separate consideration all of the terms in the left-hand side that have either C or C′ as a factor. These terms are:

$$BCv' + ABCw' + vC' + wC' = 0\ .$$

Take the left-hand side to be f(C) and we want to find f(1), which means that C = 1 and C′ = 1 - C = 0; giving f(1) = Bv′ + ABw′. We also want f(0), which means that C = 0 and C′ = 1 - C = 1; giving f(0) = v + w. With these in hand, the original equation with C eliminated is f(1)*f(0) + all of the terms that do not have C or C′ as a factor. We have f(1)*f(0) = Bv′w + ABvw′. Adding the other terms gives

$$uA' + uB' + ABu' + vB' + wA' + wB' + ABvw' + Bv'w = 0\ .$$

This equation is equivalent to the original equation and has the variable C removed. Therefore, we know the probabilities of each of the variables in this equation, other than w. This equation must be solved for w, remembering to write w′ = 1 - w. We find

$$w = \frac{uA' + uB' + ABu' + vB' + ABv}{ABv - A' - B' - Bv'}$$

We are now going to follow the procedure for Boole's Proposition III. The right-hand side of the equation for w must be expanded in terms of the sixteen possible terms made up of A, B, u, and v and their negations. In evaluating the fraction on the right-hand side of the equation it must be remembered that A′ = 1 - A and similarly for the other symbols. As Boole shows in developing his algebra of logic, the desired expansion is obtained by evaluating the fraction for all combinations of zeroes and ones for each symbol. Thus, the coefficient of ABuv in the expansion will be the numerical value of the fraction for A = B = u = v = 1. This coefficient is 1/1 = 1. The coefficient of A'Bu'v is the numerical value of the fraction when A = u = 0 and B = v = 1. Note that this simultaneously means that A' = u' = 1 and B' = v'= 0. This value is 0/1 = 0. Boole shows that the use of the various terms depends on the coefficient thus evaluated in the following way (LT, p. 92):

1. If the coefficient is 1 then the whole of the class represented by that term is to be taken.

2. If the coefficient is 0 then none of the class represented by that term is to be taken.

3. If the coefficient is 0/0 then an indefinite amount of the class represented by that term is to be taken.

4. If the coefficient is any other number then the term associated with the coefficient must be equated to zero.

In our example it will be found that ABuv is the only term with coefficient 1. There are five terms with coefficient 0: ABuv', AB'u'v', A'Bu'v, A'Bu'v', and A'B'u'v'. All of the remaining ten fall under case 4 above.

Let the sum of the terms with coefficient 1 be designated by K. Let the terms with coefficient 0 be designated L and those with coefficient 0/0 be designated M. Boole conventionally represents the fourth rule's coefficient as 1/0. Let the terms with this coefficient be designated N. Then Boole writes the general result of these manipulations of Proposition III in the form

$$w = K + 0L + (0/0)M + (1/0)\ N \quad .$$

Now, insofar as w is concerned, the logical conclusion is that w equals K plus an undetermined proportion of M. In his logical theory, Boole shows that this last equation by interpretation shows what relations exist among the elements of the original problem, independently of w. He also shows that necessarily K + L + M + N = 1. For his probability theory, instead of using N = 0 he uses K + L + M = 1 and he puts V = K + L + M and, hence, = 1. This V plays a key role in his system.

In our example, then, Boole would write:

$$w = ABuv + 0(ABuv' + AB'u'v' + A'Bu'v + A'Bu'v' + A'B'u'v') \ .$$

w, therefore, = ABuv and V = the sum of all six terms on the right-hand side of the equation. Insofar as w is concerned, the result shows that ABuv is the only one of the sixteen terms involving the events A, B, u, and v and their negations that produces w. Thus, we have indeed expressed w in terms of those events for which we know the probabilities.

This is the procedure for Proposition III. It should be clear that the algebra is somewhat heavy and that the burden rapidly increases with the number of events involved. Fortunately, as we will see shortly, this procedure can be very much simplified.

The final step required is the determination of the probability of the specified event (w in our example). This is the subject of Boole's Proposition IV and it is the essence of his method and he so indicates by his statement of the goal of Proposition IV (LT, p. 265):

> "Given the probabilities of any system of events; to determine by a general method the consequent or derived probability of any other event."

Perhaps surprisingly, there is nothing new in Proposition IV. The intellectual work has all been done and all that remains is to put the pieces together.

From the procedure followed for Proposition III we end up with

$$w = K + zM$$
and
$$V = 1,$$

where V = K + L + M. In the first of these equations z is an indefinite class symbol indicating that w includes an unknown proportion of M. In our example M = 0 so z does not enter into the analysis. Note that both V and w are the sums of terms involving A, B, u, and v, the simple events of this problem. We know the probabilities of these simple events but only subject to the condition V = 1. By Boole's assumption that compound events must be treated the same way as simple events it follows that "We must reason upon it as if experience had presented to us the events s, t, &, as simple events, free to enter into every combination, but possessing, when actually subject to the condition V, the probabilities p, q, &, respectively." (LT, p. 262)

The problem, thus formulated, is exactly the one worked out in Proposition II, discussed above. We want p′, q′, r′, and s′, the probabilities of A, B, u, and v when the restriction V is removed. The sum of all terms of V that have A as a factor is V(A) with our previous notation, and similarly for the other symbols, it follows from Proposition II that the desired probabilities will be determined by the system of equations:

$$[V(A)]'/[V]' = p \;;\quad [V(B)]'/[V]' = q$$

$$[V(u)]'/[V]' = r \;;\quad [V(v)]'/[V]' = s$$

where []′ designates the result of substituting p′, q′, r′, and s′ for the symbols A, B, u, and v, respectively. The desired probability of w is then given by

$$[K]'/[V]'.$$

Boole points out that p′, q′, . . . are determined only in order that they may be substituted in the expression for the probability of w. It is obvious, then, that we could use any other symbols having the same quantitative meaning. Therefore, Boole uses the symbols of the simple events for this purpose, in our example A, B, u, and v. This step merely involves a change in the meaning of the symbols from logical to quantitative and when Boole is about to take this step he usually says something to the effect of "passing from logic to algebra".

With this slight change in mind, the equations to be solved are:

$$V(A)/V = p \;;\; V(B)/V = q \;;\; V(u)/V = r \;;\; V(v)/V = s$$

where the resulting values are to be substituted into

$$P(w) = K/V.$$

Boole usually writes these equations differently in order to make them more symmetrical:

V(A)/p = V(B)/q = V(u)/r = V(v)/s = V

and K/x = V

where x is the desired P(w).

For our example, these equations become

$$\frac{ABuv + ABuv' + AB'u'v'}{p} = \frac{ABuv + ABuv' + A'Bu'v + A'Bu'v'}{q} =$$
$$= \frac{ABuv + ABuv'}{r} = \frac{ABuv + A'Bu'v}{s} = V =$$
$$= ABuv + ABuv' + AB'u'v' + A'Bu'v + A'Bu'v' + A'B'u'v'$$

and we want p(w) = ABuv/V.

The solution of these equations is the second major difficulty that must be overcome in applying Boole's method. The reader should really try to solve a set of them in order to appreciate the magnitude of the difficulty. I have never previously encountered relatively simple algebraic equations that are so hard to solve as these are. It is this step that justifies Boole's use of the word "repulsive" in speaking of the mathematical difficulties he had to overcome in discussing his various examples.

Boole himself finds a remarkable range of "tricks" so that he can find p(w) without separately solving for, here, A, B, u, v, and V. Since I would like to show the complete solution I did not attempt to find such a "trick". I may note that I picked this particular example out of many alternatives precisely because the solution is a relatively easy one. By this I mean that often the solution is in the form of a quadratic or cubic equation but in this example it is only a linear equation. Despite this, it took a long time to find the solution. Note that a seemingly "natural" way to try to solve the equations is to substitute a = ABuv, b = ABuv', . . ., f = A'B'u'v' and try to solve for these six unknowns plus V. However, this doesn't work because it gives seven unknowns and only five equations. Here is the solution:

$$A = \frac{p - r}{1 - q} \quad and \quad B = \frac{(q-r)(q-s)}{(q - r)(q - s) + q(1 - q - p + r)}$$
$$u = \frac{r(1 - q - p + r)}{r(1 - q - p + r) + (q - r)(p - r)} \quad and \quad v = \frac{s}{q}$$
$$V = \frac{(p-r)(q-r)(q-s)(1-p-q+r)}{(1-q)((q-r)(q-s)+q(1-q-p+r))(r(1-q-p+r)+(q-r)(p-r))}$$
$$p(w) = \frac{rs}{q}$$

A numerical example might be helpful. Using the magazine data previously quoted, we have p = .161 (magazine A), q = .229 (magazine B), r = .075 (both A and B). Let us add to these data s = .049 (both B and C) and the data is as in the above example. Since we have deduced the general formula for p(w), where w = ABC, we can calculate immediately that p(w) = .016048. However, in order to show that Boole's variables mean exactly what they are supposed to mean, here are the numerical values:

A = .111543
B = .150175
u = .795045

$$v = .213974$$
$$V = .177572$$

As we see, K = ABuv and p(w) = K/V. Here K = .00284966. Therefore, p(w) = .00284966/.177572 = .016048 as before. Similarly, any one of the six terms in V can be calculated exactly from the above values. Thus, consider A'Bu'v'. This would be the proportion exposed to B and not exposed to A and not exposed to C. This proportion is given by (1 - A)B(1 - u)(1 - v)/V = .021496/.177572 = .121048.

It is worth mentioning that there are only six terms in V because, in this example, Boole's method does not produce estimates for each of the eight logically possible classes. The discussion of Proposition III, above, ended with the logical expression for w being w = K + zM and it was noted that z is an indefinite class symbol. zM signifies that an unspecified proportion of class M is included in w. In our present example z did not occur because our w (simultaneous exposure to each of the three magazines) was completely determined. If, instead, we had taken the proportion exposed to C as our unknown we would have found that it cannot be completely determined. This would be reflected in the analysis by the occurrence of z in the logical expression for p(C). This can arise whenever V has less terms than the total logically possible.

One important aspect of the present example deserves heavy emphasis. This is that it is exactly like the Challenge Problem in that there is no possible way to estimate the proportion exposed to all three magazines with the methods of orthodox probability theory. The present example is actually very similar to the Challenge Problem. Formulating the Challenge Problem in terms of magazine data, we need to be given p(A), p(B), p(AC) and p(BC) where C is the event whose probability is desired (E in the presentation of Section 1). Thus, the only difference between this example and the Challenge Problem is that we are given p(AB) here instead of p(AC) in the Challenge Problem. All of the denials and criticisms of Cayley, Wilbraham, Macfarlane, MacColl, and Keynes apply equally to this example. While the underlying argument will be made very much clearer subsequently, the reader might already wonder about the relevance of the various objections to Boole's actual argument. What has Boole's definition of independence got to do with any step in the procedure we have just completed? in fact, nothing whatever. Boole's method may be subject to criticism or even rejection but, if so, it has nothing whatsoever to do with the historical criticisms of it.

The mathematical difficulties caused Boole much trouble in presenting his method in An Investigation of the Laws of Thought. He demonstrates remarkable ingenuity in finding ways to overcome the algebraic complexities but the solutions he usually obtained were too involved to permit him to make the kinds of general statements he clearly wanted to make or even to verify the solution by common sense. There were no calculators or computers and the arithmetic was ordinarily too onerous for pencil-and-paper calculations. I have no doubt that these problems seriously affected the reception accorded Boole's method and certainly adversely affected Boole's own understanding of his method.

Today, with the easy availability of computers and numerical methods, I would argue that even if Boole's method were as complex and difficult as it appears to be, it would still demand attention because it permits the solution of important probability problems that cannot be solved otherwise. However, as we will now see, his system admits of remarkable simplification with the result that many problems, including most of his own examples, can be quite easily solved even with pencil and paper.

Section 14. Boole's Method: A Verbal Transcription

Suppose we have a probability problem involving some number of random events, say x, y, z, . . Some of these events may be ones that we would call simple and some may be ones that we would call compound. We would certainly call an event compound if it were defined in terms of some of the other events, for example, z = xy. Boole assumes that this distinction is not an important one. If some of the events are compound, Boole intends to designate each of them by a single letter and then to treat them as if they were simple.

We can, therefore, assume that the probability problem involves some number of simple events x, y, z, . . . Our desideratum is to find the probability of some event, w, that is defined in terms of one or more of the simple events. If w were not so defined there would be no relationship between the data of the problem and the desired answer, p(w).

We will know the probabilities of some, perhaps all, of these simple events. The first step in Boole's procedure is to use his algebra of logic to eliminate all of the variables representing simple events for which the probabilities are not known and then to express the event w in terms of the simple events for which the probabilities are known. This step is the goal of Proposition III.

Boole makes the assumption that if random events x and y with probabilities p and q are known and nothing is known about the probability of the joint occurrence of x and y, then we assume p(xy) = pq. What might be known about p(xy)? First, we might know that p(xy) = r, say. In this case we would have defined an event t = xy and we would have included t as a simple event with known probability, p(t) = r. Second, we might know that xy cannot occur, p(xy) = 0, if either x or y occurs or that xy must occur, p(xy) = 1, if either x or y occurs. In this case Boole would call the events x and y conditioned. Any such information must be expressed as a logical equation and included in the system before the analysis of the previous paragraph. Thus, if xy cannot occur if either x or y occurs then the logical equation would be xy = 0. If xy must occur if either x or y occurs then the logical equation would be xy' + x'y = 0.

The result of the logical analysis associated with Proposition III is the determination of all of the combinations of the simple events with known probabilities that are logically possible. Boole designates the sum of these terms as V. Boole now argues that the known probabilities of the simple events are, in fact, obtained under the condition V = 1, meaning simply that V is true. This is necessarily true since V includes all logically possible combinations of the simple events with known probabilities and their negations. Boole wishes to determine what the probabilities of these simple events must be without the condition V = 1 in order that the probabilities that we have observed (the known values) would be what they are. This step leads to the formulation of the algebraic equations, exemplified in Section 13, that are often very difficult to solve. The whole argument here is essentially Boole's Proposition IV.

Let [V] designate the arithmetical value of V when the solved probabilities are substituted into V. Let V(w) be the terms in V that are included in w and let [V(w)] be the arithmetical value of V(w) when the solved probabilities are substituted into V(w). Then the desired probability is

P(w) = [V(w)]/[V} .

This is Boole's solution to the general problem of probability theory as he has formulated it.

The example of Section 13 (and the example in Appendix II) should suffice to show that Boole's own procedures for accomplishing this analysis were onerous. Keynes was justified in referring to the extreme difficulty of Boole's analysis and, as we referenced earlier, even Boole himself referred to some of his examples as "repulsive" because of their algebraic complexity. Fortunately, his analysis can be very much simplified.

It is clear from the two examples (those of Section 13 and of Appendix II) that considerable labor can be required to follow Boole's logical algebra in order to expand the unknown, w, as a function of the events with known probabilities. This step can be made trivially easy and this simplification will be discussed in Section 15.

The other major difficulty in accomplishing Boole's analysis is the solution of the equations that his method produces. While this step cannot be made "trivially" easy it can be reduced to a very straightforward process which is a vast improvement. The resulting procedure makes the whole analysis much more perspicuous. This approach will be presented in Section 17.

The intermediate Section 15 will be devoted to a discussion of the main single conceptual step in Boole's analysis. This is the argument that leads from the logical expansion of w to the equations. Perhaps some discussion of the reasoning that underlies this step will make it clearer. It is surely worth trying to clarify this step because it is the essence of Boole's method. However, Section 15 will provide only a preliminary look at the structure of Boole's thinking in this step. A more detailed statement will be made subsequently.

Section 15. The Determination of Boole's V

Here is a quick outline of Boole's procedure for obtaining his V. Suppose there are simple events designated by x, y, z, . . . and S, T, . . . are compound events whose probabilities are known and which are known functions of x, y, z, . . . Represent S, T, . . . by single letters s, t, . . . and thus as simple events. So doing will produce a series of logical equations connecting the symbols s, t, . . . with the symbols x, y, z, . . . Let W represent any event whose probability is desired. Obviously, then, W is also a known function of x, y, z, . . . Similarly, represent W by a single logical symbol, w. The resulting system of equations can be converted to one equation and from this equation any of the symbols x, y, z, . . ., and perhaps some of the symbols s, t, . . ., that represent events with unknown probabilities are to be eliminated. Finally, the resulting equation is solved for w, using Boole's logical algebra. This requires expanding w in terms of the remaining symbols and the result is Boole's V.

It needs to be emphasized that once V is available the formulation of the equations that have to be solved is immediate. All that is needed are the quantities we have previously designated as, for example, V(x), where V(x) stands for all the terms in V that have x as a factor (and not including terms that have x' = (1 - x) as a factor).

Boole himself notes that V as he defines it consists of "all the possible constituents which can be formed from the symbols s, t, &, i.e. from all the symbols representing events whose probabilities are given". (LT, p. 263) However, he does not seem to have noticed that this means that we can write down all the constituents that occur in V almost immediately, using only simple Boolean multiplication. There is no need to go through the laborious process of expansion. Thus, consider the example of Section 13, where we are given p(A), p(B), p(AB), and p(BC). The simple events for which we know the probabilities are A and B. Designate the compound events by u = AB and v = BC. V will have all logically possible terms involving A, B, u, and v. But any term involving u and/or v is exactly equivalent to some term involving only A, B, and C and we can reason in terms of all possible constituents based on A, B, and C with no loss of information. Thus, for example, $u'v'AB' = AB'$. We will lessen the difficulties considerably by reasoning wherever possible with terms involving A, B, and C rather than A, B, u, and v.

By elementary Boolean algebra we have that $u' = A' + B'$ and $v' = B' + C'$. Therefore, doing the multiplications in two steps for convenience:

$$uv = ABC$$
$$uv' = ABC'$$
$$u'v = A'BC$$
$$u'v' = A'B' + A'C' + B'C' + B'$$

These can be written down almost at sight, as can be the results of multiplying these by AB, AB′, A′B, and A′B′:

$uvAB = ABC$	$u'v'AB' = AB'$
$uv'AB = ABC'$	$u'v'A'B = A'BC'$
$u'vA'B = A'BC$	$u'v'A'B' = A'B'$

All of the other terms are impossible. The only logical rule used here is that the product of any two terms involving, say, A and A′ is logically impossible. Note that $AB' = AB'C + AB'C'$ and similarly for A′B′. Thus, there are eight terms on the right-hand sides of the equations. These are all the possible terms made up of A, B, C, and their negations.

This simple alternative procedure is exactly equivalent to Boole's argument and, as we will see, is enormously easier. This method of getting V is so quick and easy that it is somewhat surprising that Boole overlooked it. Probably the reason for this is that Boole was concentrating his attention on the logical procedure for obtaining w in terms of the events with known probabilities. So doing, he did not notice that he could get the expression for w much more quickly after obtaining the expression for V. Thus, in our example $s = ABC$ so $V(w)$ is simply ABC. Hence uvAB, reading off the left-hand side of the equation.

This very much quicker method of getting V has a corollary advantage in that it does not, as Boole's procedure does, suggest that there is anything particularly important about the specific event for which the probability is desired, w. In our example, suppose w were defined as A′BC′. Then we have immediately that $V(w) = u'v'A'B$, as the left-hand side of the equation with A′BC′ on the right. In other words, however w might be defined we can determine V(w) directly, and almost instantaneously, by finding the left-hand side of the appropriate equation in our table above. Thus, suppose that w were defined as $w = B(AC + A'C')$. Then it is easy to see that $V(w) = uvAB + u'v'A'B$.

We will find that further important simplifications of Boole's argument result from paying primary attention to constituents based on the original simple events (A, B, and C in our example). This way of finding V will be repeatedly illustrated in the subsequent examples. For the present it is sufficient to emphasize that it converts what is otherwise a very lengthy algebraic manipulation into an extremely quick and easy procedure.

Section 16. The Argument from V to the Equations

As stated earlier, the most distinctive feature of Boole's method is the argument he used to deduce, from the expression for V, equations that determine the desired unknown probability. Boole's exposition is by no means clear and the argument is a curiously slippery one. Therefore, in this Section I will offer an illustration of how the argument works. An indication that this is not easy to do is suggested by the fact that it is precisely here that Boole had to add the lengthy Note that precedes the title page of An Investigation of the Laws of Thought. Apparently, he concluded while his book was in press that his own example to illustrate this argument was faulty.[11] Perhaps this was because Wilbraham, in his article attacking Boole's solution of the Challenge Problem, specifically criticized the example in question.[12] (Note that this example of Boole's will be discussed at length subsequently.) I believe that Boole's original solution was completely correct and that he made his mistake when he added the Note. But in any case, this surely suggests that it is necessary to proceed cautiously. And we might note that here again, as with the Challenge Problem, Boole had remarkably bad luck in his selection of key examples.

In order better to follow the operation of Boole's procedure, let us consider an example where his approach is not needed. Suppose we have complete information about two random events, A and B. Suppose that P(A) = p, P(B) = q, and P(AB) = r. Clearly, there is nothing to estimate here because we know all the probabilities. Thus, there are four possible combinations of A and B and their negations and we know the probabilities of each of them:

$$P(AB) = r; \qquad P(AB') = p - r; \qquad P(A'B) = q - r; \qquad P(A'B') = 1 - p - q - r \quad .$$

Nonetheless, suppose we follow Boole's procedure. Set t = AB so t′ = A′ + B′ and by simple logical multiplication we have:

$$tAB = AB; \qquad t'AB' = AB'; \qquad t'A'B = A'B; \qquad t'A'B' = A'B' \quad .$$

All the other combinations of t, A, and B and their negations are logically impossible. Therefore,

$$V = tAB + t'AB' + t'A'B + t'A'B'$$

and the equations are

$$\frac{tAB}{r} = \frac{tAB + t'AB'}{p} = \frac{tAB + t'A'B}{q} = V$$

There are three equations in three unknowns. The solution is, now letting, as Boole does, A represent P(A), and similarly:

$$A = \frac{p - r}{1 - q} \quad and \quad B = \frac{q - r}{1 - p}$$

$$t = \frac{r(1 - p - q + r)}{r(1 - p - q + r) + (p - r)(q - r)}$$

The specific values of these Boolean variables are really not of any importance at all. Their only purpose is to exactly reproduce the probabilities that are known in a problem under Boole's two basic assumptions: compound events must be treated as if they were simple and events without known joint probabilities must be treated as if they were independent. The solution given here does this for the present example. Here is a numerical illustration. Suppose p = .4, q = .3, and r = .2. Then we can calculate that A = 2/7, B = 1/6, and t = 5/6. From these values it follows that V = 25/126. Then all the given probabilities satisfy the equations exactly. Thus, tAB/V = (5/6)(2/7)(1/6)/(25/126) = .2, as it must, and similarly for the other given probabilities.

Now suppose, in this simple example, that we do not know P(AB). Then, evidently V = AB + AB′ + A′B + A′B′ and the equations are the solution of these two equations in two unknowns is

$$A = p \quad \text{and} \quad B = q \ .$$

Here one might want to estimate the unknown p(AB) and the estimate is immediately just pq. This is in accord with the basic assumption: if all we know are the probabilities of A and B then we assume that they are independent in estimating the joint probability.

As stated above, the specific values of the Boolean variables are essentially of no independent interest. In fact, in the majority of the solutions of examples Boole gives in An Investigation of the Laws of Thought, he succeeds in finding some way (often a “trick”) of solving for the desired probability without explicitly finding the values of the intermediate variables at all. This fact will be of some importance subsequently.

Now, while the easy example of this Section has only two simple events it is typical of the general situation. With complete information about all probabilities based on n simple events there will be 2^n - 1 Boolean variables and their values will exactly reproduce the known probabilities. If only the probabilities of the n simple events are known there will be n Boolean variables and their values will be the known probabilities. As a result, the estimated probability of any specified compound event based on the known simple events will be the probability calculated under the assumption that the simple events are independent. These are, so to speak, the two extremes. It seems plausible that in intermediate cases the same method ought to give reasonable estimates of unknown probabilities.

Section 17. The Solution of the Equations

The difficulties of solving the equations can be made very much easier by using a simple, and obvious, device to summarize the information we have about the problem under consideration. Subsequently we will use the same approach to obtain an exact characterization of Boole’s method.

I will discuss the device in the context of probability problems involving three simple events. In Section 20 we will examine some of Boole’s examples based on just two simple events but most of his examples, including the Challenge Problem, are based on three simple events. Needless to say, the elementary device to be presented here can be used for any number of simple events.

Therefore, let the three simple events by A, B, and C. If we had complete information about these three events it could be presented in a table like this:

	A		A′	
	B	B′	B	B′
C	a	b	c	d
C′	e	f	g	h

Here p(ABC) = a, p(A′BC′) = g, and so forth. Obviously, the sum of the eight entries in the body of the table must equal one.

The advantage of using this table is that any information that is known about the problem (thus, any specified probabilities) can be summarized by an appropriate equation involving one or more of the eight entries in the table. This will permit writing Boole's equations much more simply and it also makes it straightforward to solve them.

Let us illustrate the procedure on the example used in Section 13. It will be recalled that in that example we assumed that p(A) = p, p(B) = q, p(AB) = r, and p(BC) = s were known values. This means that the following equations are true:

$$
\begin{aligned}
a + b + e + f &= p \\
a + c + e + g &= q \\
a + e &= r \\
a + c &= s \\
a + b + c + d + e + f + g + h &= 1
\end{aligned}
$$

Since there are five equations in eight unknowns we can solve for five unknowns in terms of three others. "a" is an obvious first choice for the right-hand side and one quickly obtains:

$$
\begin{aligned}
e &= r - a \\
c &= s - a \\
b &= p - r - f \\
g &= q - s - r + a \\
d &= 1 - p - q + r - h
\end{aligned}
$$

Since "f" and "h" occur on the right-hand side in only one equation each, it is evident that we cannot solve for either of them and we can equally well write the equations as:

$$
\text{(I)} \qquad
\begin{aligned}
e &= r - a \\
c &= s - a \\
b + f &= p - r \\
g &= q - s - r + a \\
d + h &= 1 - p - q + r
\end{aligned}
$$

The equations with two unknowns on the left-hand side show us that there is no way we can separately estimate the two unknowns. Thus, any solution will have b + f = p - r and we cannot determine "b" and "f" separately.

In Section 14 we already obtained V by a simplified procedure involving only logical multiplication. We found the following, where we include C and C′ on the right-hand sides even though C is not a Boolean variable (since we do not know p(C)):

$$
\begin{aligned}
uvAB &= ABC \\
uv'AB &= ABC' \\
u'vA'B &= A'BC \\
u'v'AB' &= AB'C + ABC' \\
u'v'A'B &= A'BC' \\
u'v'A'B' &= A'B'C + A'B'C'
\end{aligned}
$$

It will be noted that all of the possible combinations of A, B, C, and their negations are included on the right-hand sides of these equations. Each one of these combinations is represented by one of the letters "a", . . . , "h" in the table. Therefore, since we are going to proceed to using arithmetical values, we can write the equations thus:

$$
\begin{aligned}
uvAB &= a \\
uv'AB &= e \\
u'vA'B &= c \\
u'v'AB' &= b + f \\
u'v'A'B &= g \\
u'v'A'B' &= d + h
\end{aligned}
$$

It will be seen that these equations reflect exactly the same fact that we discovered above, namely that "b" and "f" cannot be separately determined, nor can "d" and "h".

In Section 13, following Boole's own procedure, we ended up with four equations involving the four Boolean variables, u, v, A, and B. Here we have the same unknowns but by Boole's same reasoning we now have six equations:

$$\frac{uvAB}{a} = \frac{uv'AB}{e} = \frac{u'vA'B}{c} = \frac{u'v'AB'}{b+f} = \frac{u'v'A'B}{g} = \frac{u'v'A'B'}{d+h} = V$$

Using the present approach the equations will always have this general form. Appendix C gives a general procedure for solving this kind of equations. A convenient way is to divide all the equations through by uvAB and then set u″ = u′/u, v″ = v′/v, A″ = A′/A, and B″ = B′/B. We have more equations than we need so we can forego dividing V by uvAB. We get:

$$\frac{1}{a} = \frac{v''}{e} = \frac{u''A''}{c} = \frac{u''v''B''}{b+f} = \frac{u''v''A''}{g} = \frac{u''v''A''B''}{d+h}$$

We know from equations I that we really have only one unknown, "a", so we need to find only one equation. We will get this equation by finding any one of the Boolean variables that can be written in two different ways in terms of "a", . . . , "h". In other words, we want to find two different ways of obtaining the same numerator, usually by multiplying two or more of the fractions together. A moment's inspection shows that the product of the 2nd and 3rd of these expressions has the same numerator as the product of the 1st and 5th:

$$\frac{u''v''A''}{ec} = \frac{u''v''A''}{ag}$$

Hence, ec = ag. Substituting the expressions from i gives

$$a(q - s - r + a) = (s - a)(r - a)$$

and, immediately

$$a = \frac{rs}{q}$$

This, of course, is the same answer as before. The reader will not thoroughly appreciate the literally enormous improvement in the ease of finding the solution unless he or she undertakes at least once to solve the equations in Boole's form. It is surprisingly difficult. Of course, it is trivially easy to show that the form of equations given above is equivalent to the ones deduced by Boole.

Given "a" it follows that the logical table of A, B, and C can be immediately reconstructed as far as this is possible (since "b" and "f" cannot be separately determined, nor can "d" and "h"). This is done simply by substituting "a" into the equations (I). Writing the possibilities in terms of A, B, and C, we calculate, from equations (I) that

ABC = .016048	AB' = .086
ABC' = .058952	A'BC' = .121048
A'BC = .032952	A'B' = .685

It is also worthy of emphasis that it is not at all necessary to determine the numerical values of the Boolean variables although, of course, they can be found if desired.

This example shows that the approach presented here makes reasonably easy that which would otherwise be quite difficult. The idea is so simple and the gain is so great that it is perhaps surprising that Boole overlooked it. Nonetheless, I think that it is indubitable that he did overlook it. There are three places in An Investigation of the Laws of Thought where Boole's statements at least implicitly suggest that he was not aware of the advantages of using a simple logical table in the way shown here. Since I am quite certain that this accounts for some deficiencies in Boole's method, it is worth a brief digression to demonstrate it.

The construction of the logical table associated with a given set of classes seems to me to far and away the most straightforward approach to Boolean logic. Martin Gardner has given us a fascinating account of the variety of logical diagrams that were proposed to help visualize the relationships between logical classes.[13] Euler, Venn, and Lewis Carroll are a few of the persons who attempted to find visual representations of logical relationships. With all due respect to these illustrious persons, many years of teaching the subject have shown me that no other approach is nearly so easy to use and understand as an ordinary logical table. This way of thinking about Boolean logic is philosophically satisfying because it leads to the insight that any and every logical relationship in Boolean algebra can be correctly interpreted as the denial of existence of one of the logical classes in such a table.

Boole certainly did not recognize that any such device would be useful. In this context his Example 5 of Chapter IX is interesting (LT, pp. 146-9). This Chapter is entitled "On Methods of Abbreviation" and is Boole's attempt to find ways to partially mitigate the difficulties of his logical calculations. Example 5 has five classes and, after considerable algebra, Boole gives his desired conclusions. The conclusions are a little complicated and Boole's final statement is (LT, p. 149) "I have not attempted to verify these conclusions." My present point is not that the use of the logical table makes the whole analysis very much quicker and easier, although that is certainly true. Rather, having the table makes it trivially easy to verify Boole's conclusions. The fact that he thought it would be difficult to verify his conclusions is sufficient evidence that he was not aware of the usefulness of the logical table.

The other two examples that indicate that Boole was unaware of the usefulness of the ordinary logical table both come from his examples on probability theory. The first of these is his Example 3 of Chapter XVIII (LT, pp. 279-81). This example is the one where Boole clearly made an erroneous statement. Designate A as being that witness A speaks the truth, with p(A) = p, and B as being that witness B speaks the truth, with p(B) = q, and the probability that they disagree is r. Thus, there are three known probabilities. The third one is p(AB′ + A′B) = r. Boole asks for the probability that if they agree, their statement is true. This is an instance where there are only two simple events and the logical table is:

	A	A′
B	a	b
B′	c	d

With the given information we have the equations:

$$a + c = p$$
$$a + b = q$$
$$b + c = r$$
$$a + b + c + d = 1$$

There are four equations in four unknowns and the solution is immediately:

$$a = (p + q - r)/2$$
$$b = (q + r - p)/2$$
$$c = (p + r - q)/2$$
$$d = (2 - p - q - r)/2 .$$

The answer to the question Boole asked is a/(a + d) or (p + q - r)/2(1 - r). Boole's method is not even needed for this problem. Nonetheless, Boole goes on to state concerning his solution (LT, p. 281): "It is probable, from the simplicity of the results (5) and (6), that they might have been deduced by the application of known principles; but it is to be remarked that they do not fall within the scope of known methods. The number of the data exceed that of the simple events which they involve." Macfarlane, in his book, pointed out this error. It seems reasonable that the only way Boole could have made such an elementary blunder was by thinking that there is just one variable for each simple event. In other words, he was not thinking in terms of the underlying logical table.

The third example is Example 4 of the same Chapter (LT, pp. 281-4). This one can be summarized quickly. Boole deduces his desired solution and then wishes to verify it in particular cases. He does this for one such case and then states (LT, p. 284): "Perhaps the general verification of the above solution would be difficult." The present point is that Boole could only think this because he had only deduced one specific probability in the logical table. Under this circumstance it is difficult to verify the solution. However, if he had deduced the probabilities for each of the logical possibilities he would have seen that it is very easy to verify the solution. Once again, it seems clear that Boole did not think in terms of the underlying logical table.

Section 18. Boole's Procedure: Simplified and Summarized

We can simplify and summarize Boole's procedure so that most typical problems can be handled in a more-or-less routine manner. There are always some more complicated examples that will not fit into such a generalized scheme but Boole's own examples do fit. Here is a sequence of steps that usually need to be accomplished in the order given.

1. A simple event is one that initially and by ordinary usage would be represented by one letter in a symbolic representation. A compound event is one that initially and by ordinary usage would be represented by two or more letters in a symbolic representation. This means that a compound event involves two or more simple events.

2. Represent all simple events by appropriate single letters and represent all compound events by appropriate combinations of individual letters for the simple events involved. Thus, the letters used might be A, B, C, . . .

3. List all the events, simple or compound, for which there are known probabilities. Suppose there are m such events. For these events, list all the letters that occur in their symbolic representation. Suppose that n such letters occur.

4. Use the list of n letters from (3) to form a logical table showing every possible combination of these letters and their negations. If there are n such letters then there will be 2^n possible combinations of them. Represent each such possible combination (in the table) by an appropriate symbol, perhaps

a, b, c, . . .

5. Each of the m known probabilities can now be represented as a linear equation in terms of a, b, c, . . . Write these equations. If there are m known probabilities then m of the unknowns (a, b, c, . . .) can be solved for in terms of the others. (NOTE THAT AT THIS POINT YOU MAY BE ABLE TO SOLVE THE GIVEN PROBLEM DIRECTLY FROM THE EQUATIONS. IF THIS IS TRUE THEN YOU DO NOT NEED BOOLE'S PROCEDURE).

6. Designate each compound event by a single letter except that the desideratum does not have to be so designated, even if it is a compound event. As a result of this step the m events with known probabilities will be designated by single letters, where some of the letters may be A, B, C, . . . for simple events and others will be, say, W, X, Y, Z, . . . for compound events.

7. For the list of m events with known probabilities there will be 2m possible compound events, each one involving all m of the letters assigned in (6) or their negations. We want to know which ones of these compound events are, in fact, logically possible. Using the definitions of these compound events in terms of A, B, C, . . ., use Boolean logical multiplication to express the combinations in terms of A, B, C, . . . Only those compound events that occur in the various logical multiplications are, in fact, logically

possible. Any conditioning of events is taken into account here. By conditioning I mean, as did Boole, that some logically possible event is prohibited from happening or, perhaps, is required to happen. It does not mean what is ordinarily intended when speaking of conditional probabilities. The sum of the logically possible terms found in this step is V.

8. Each logically possible compound event of (7) is, as a result of (6), expressed in terms of A, B, C, . . . and, hence, occurs in the logical table found in (4). Therefore, each logically possible compound event of (7) can be equated to the sum of one or more terms a, b, c, . . . In the logical table of (4). This gives equations of the form: AB'WX = a, A'W'X'Y = b + c, and so forth. Generally, represent these equations as L(i) = R(i), where L(i) will be like AB'WX and R(i) will be like a, or b + c.

9. The set of Boolean equations that must be solved, then, is given immediately by

$$\frac{L(1)}{R(1)} = \frac{L(2)}{R(2)} = \frac{L(3)}{R(3)} = \ldots = V$$

10. The Boolean variables in these equations are the m single letters used to represent each event, whether simple or compound, as in (6). However, it is not at all necessary to solve for these as the unknowns. The unknowns for which solution is needed are the letters a, b, c, . . . occurring in the denominators of the fractions in (9). It is usually quite easy to solve for these directly.

11. The solution is complete. All of a, b, c, . . . may now be expressed in terms of the known probabilities. If this is not true it is the case that as many of them are so expressed as is possible granted the original data of the problem. Any desired probability can be calculated directly from the logical table of (4). Such a probability may, of course, not be completely determined. It should be explicitly noted that if the desired probability cannot be represented in terms of a, b, c, . . . then it cannot be found from the data. In particular, these statements also refer to the probability of the desideratum. Knowledge of the values of a, b, c, . . . , or as many of them as can be calculated, is essentially complete knowledge, given the original known probabilities. Whatever the definition of w, the desired probability of w, P(w), can be written out in terms of a, b, c, . . . , thus solving the stated problem.

This is a step-wise procedure for accomplishing Boole's analysis of probability problems but using the two major simplifications presented earlier.

It should be emphasized that this is entirely Boole's own analysis. The analysis is based on Boole's assumptions, previously discussed. The most essential one is that either the joint probability of two events, whether the events are simple or compound, is known and, hence, included in the data of the problem or it is not known and, therefore, the joint probability is calculated under the assumption that the two events are independent.

The major difference between the step-wise summary and Boole's procedure is in Step 9, where the equation to be solved does not have a term of the form, (A + cC)/u, where u represents the desired p(w). (For a general example, see (12), LT, p. 267.) The reason is that our approach provides us with the whole logical table, as completely filled out as is possible with the given information. Therefore, we can determine p(w) directly from the table and we do not have to solve for it. The effect of this change is a very great decrease in the algebraic manipulations that are required. We will give examples of how this is done in Sections 20 and 21 on some of Boole's examples.

Section 19. An Example Comparing Boole's Method With the Proposed Emendations

It may be helpful to have an example that, on the one hand, shows how Boole's own procedure works and, on the other hand, explicitly compares that procedure to the argument that results when the analytical alternatives presented in this book are used. For this purpose, let us consider an example with three simple events for which we know p(A) = p, p(B) = q, p(C) = r, and, in addition, we know the probabilities of two complex events, p(AB) = s, and p(BC) = t. This means that we do not know p(AC), nor do we know p(ABC).

We will consider Boole's own procedure first. It is important to emphasize that there is already a major, and important, difference between Boole's argument and the proposed emendations. Boole's whole method is based on the necessity of replacing complex events by simple ones and we must equally well do this for our emended method. Since we know the probabilities of two complex events Boole would let s = AB, so we know p(s) = u, and t = BC, so we know p(t) = v, and we must do likewise in our emended method. The "important difference" referred to is that with Boole's procedure we need to similarly convert a complex desideratum to a simple event. Thus, suppose Boole's desideratum is the complex event, ABC. Then he would convert this to a simple event, x = ABC, and explicitly calculate the resulting p(x) = w, say. If he subsequently wanted to find p(AC), then, he would have to repeat the whole analytical procedure. In his book there are various examples for which the necessity of repeating the analysis leads to some lengthy algebraic calculations. With the emended method this does not occur. Our goal is to complete, within the limitations of our available knowledge, the entire probability table for the three simple events, including all four of the complex events that are defined in terms of the given three simple events.

Therefore, in accordance with the preceding paragraph we will specify that Boole wants to find w = p(x) where x = ABC. Then the information available takes the form of three equations: s = AB, t = BC, and x = ABC. Following Boole, we negate each of these equations, giving three equations each equal to zero: s'AB + sA' + sB' = 0, t'BC + tB' + tC' = 0, w'ABC + wA' + wB' + wC' = 0. (in these equations, s' = (1 - s), t' = (1 - t), and w' = (1 - w).) We are entitled to add these three equations together and thus obtain one equation:

$$s'AB + sA' + sB' + t'BC + tB' + tC' + w'ABC + wA' + wB' + wC' = 0.$$

(At this point in the analysis it may be the case that we do not know the probabilities of one or more of the simple events. This is not the case for our example but, as an illustration of this possibility, suppose that we do not know p(C). Then we could use a simple, and standard, manipulation of Boolean logic to eliminate all of the terms involving C and C' = 1 - C. The new equation would express the desideratum in terms only of those events for which we know the probabilities.) The parenthetical possibility is not true for our example so we follow Boole and solve the above equation for w, making sure that we remember to include w' = 1 - w. This gives an expression for w in terms only of events with known probabilities:

$$w = \frac{s'AB + sA' + sB' + t'BC + tB' + tC' + ABC}{ABC - A' - B' - C'}$$

Following Boole, we now wish to express was a developed function of the five variables: s, t, A, B, and C. There are two steps involved. First, we must recognize all the possible constituents of such a development and, second, we must calculate the coefficients of each of the constituents because in Boole's system the coefficients will tell us the role played in the development by the corresponding constituent. It is easy to list all of the possible constituents. There are five variables and each variable can assume a value of either 1 or 0 so there are 25 = 32 constituents. The coefficient of a given constituent, say stABC, will be given by evaluating the expression for w, using 1 for each variable which occurs as s, t, A, B, or C and using 0 for each variable that occurs as s′, t′, A′, B′, or C′. For the constituent stABC it will be found that the numerator of w is 1 and the denominator is also 1 so the coefficient is 1. This means that this constituent, stABC, will certainly include the desideratum, x = ABC.

Calculating the coefficients for each of the 32 constituents is fairly lengthy for pencil-and-paper calculations and the procedure could be quite easily programmed for computers if it were to be done frequently. This fact recognized, it is nonetheless fairly straightforward to make the work somewhat tolerable by using a tabular approach. Specifically, let us proceed with a table that includes all the constituents and permits the relatively quick calculation of the numerators and denominators, and hence the coefficient, for each constituent. Remember that in a row with s = 1 it follows that s′ = 0 and if

A = 0 then A′ = 1, and similarly for the other variables:

0	1	0	1	1	0	-1	0	s′tA′BC
0	0	0	0	1	0	-2	0	s′t′A′B′C
0	0	0	0	0	0	-3	0	s′t′A′B′C′
0	0	1	0	1	0	-1	0	s′t′AB′C
0	0	1	0	0	0	-2	0	s′t′AB′C′
0	0	0	1	0	0	-2	0	s′t′A′BC′

The column headed N gives the values of the numerators and D shows the values of the denominators.

This table is not yet completed. Half of the constituents are shown - only those with s = 1. Boole shows that the coefficients can only be 1, 0, 0/0, or any other number, which Boole conventionally writes as 1/0. He also shows that the use of the various constituents depends on the coefficient in the following ways: 1. If the coefficient is 1 then the whole of the class represented by that term is to be taken; 2. If the coefficient is 0 then none of the class represented by that term is to be taken; 3. If the coefficient is 0/0 then an indefinite amount (none, some, all) of the class represented by that term is to be taken; and

4. If the coefficient is any other number (conventionally, Boole writes it as 1/0) then the term associated with the coefficient must be equated to zero. Boole writes the general result of these manipulations in the form

$$W = K + 0L + (0/0)M .$$

In sofar as w is concerned, the logical conclusion is that w equals K plus an undetermined proportion of M. In his probability theory Boole uses K + L + M = 1 and he puts V = K + L + M = 1. Instead of completing the entire table I will show the constituents which have coefficients of 0:

0	1	0	1	1	0	-1	0	s′tA′BC
0	0	0	0	1	0	-2	0	s′t′A′B′C
0	0	0	0	0	0	-3	0	s′t′A′B′C′
0	0	1	0	1	0	-1	0	s′t′AB′C
0	0	1	0	0	0	-2	0	s′t′AB′C′
0	0	0	1	0	0	-2	0	s′t′A′BC′

There is only one constituent with a coefficient of 1. This means that there is only one of the 32 terms involving the events s, t, A, B, and C and their negations that produces w. There are no constituents with coefficient 0/0. Therefore, Boole would now write that w = stABC + 0(st'ABC' + ... + st'A'BC') and that V = sum of all of the eight terms on the right-hand side of the equation.

Boole is now ready to write the equations that need to be solved in order to find the desired value of x = p(w). Boole lets V(A) represent all of the eight constituents that include A, V(B) similarly if B is included, V(C) likewise if C is included, V(s) if s is included, and V(t) if t is included. Then the equations are:

$$\frac{V(A)}{p} = \frac{V(B)}{q} = \frac{V(C)}{r} = \frac{V(s)}{u} = \frac{V(t)}{v} = V$$

where the resulting values are to be substituted into K/x = V, giving the desired x = p(w). Writing these equations out, they are:

$$\frac{stABC + st'ABC' + s't'AB'C + s't'AB'C'}{p} + \frac{stABC + st'ABC' + s'tA'BC + s't'A'BC'}{q}$$
$$+ \frac{stABC + s'tA'BC + s't'A'B'C + s't'AB'C}{r} + \frac{stABC + st'ABC'}{u} + \frac{stABC + s'tA'BC}{v}$$
$$= V = stABC + \ldots + st'A'BC'$$

and we want x = p(w) = stABC/V. V is still the sum of the eight constituents included in this equation. In other words, using Boole's notation we have V = V(A) + V(B) + V(C) + V(s) + V(t).

Solving a set of equations such as these is a thoroughly tiresome algebraic exercise. I know of no "good" way to attack such equations and I don't think Boole knew one either. I am quite certain that it was the difficulty of solving such sets of equations that led Boole to refer to obtaining such solutions as being "repulsive". Further, it is a fact easily verified that Boole never gives any hint as to how to proceed in solving such equations. All of this granted, it obviously remains the case that such a set of equations can be solved. What suggestions can I make about a procedure to be followed?

The most straightforward approach seems to be finding the values of the Boolean variables. These are all of the variables that occur in the numerators of the fractions. For our example, the Boolean variables are s, t, A, B, and C and their complements, s' = 1 - s, t' = 1 - t, and similarly for the other three variables. To illustrate the idea, suppose we are analyzing a specific instance of the general structure of our example. In particular, suppose that for our specific example we have p(A) = .3, p(B) = .4, p(C) = .5, p(s) = .18, and p(t) = .28. Suppose, finally, that we have found the values of the Boolean variables to be: A = .2, B = .178378, C = .366667, u = .765957, and v = .801205. Now consider the final fraction in our equation:

$$\frac{stABC + s'tA'BC}{v} = V$$

Using the values of the Boolean variables we can calculate that stABC = .0080027 and s'tA'BC = .0098117. Remembering that v = p(t) = .28 we calculate that V = .0637119 and, incidentally, and for example, we immediately can find p(w) = stABC/V = .126.

This shows that having the values of the Boolean variables can be very useful and, indeed, can amount to solving the equations. How, then, should we find the values of the Boolean variables? Three comments are needed. 1. It is far too difficult to find the values of the Boolean variables directly from the equations. 2. The way we write the equations in our emended method makes it quite easy to find the values of the Boolean variables so we will show how to find them when, very shortly, we turn to a discussion of the emended method. 3. Even granted the greater ease of finding the values by our emended method, it still remains true that it is only relatively easy when we are dealing with a specific numerical example; if we have symbolic values in our example instead of numerical ones then the work involved becomes prohibitive. Note that this does not occur when we use the whole emended method itself. The advantages of the emended method are, to put it bluntly, enormous.

We will now turn to a consideration of our proposed emendations to Boole's method. Everything follows from recognition of the fact that there are very major advantages to be derived from constructing a complete logical table for the initial number of simple events there are in the problem we are considering. Thus, the example we have been using in this Section has three initial simple events: A, B, and C. Therefore, there will be $2^3 = 8$ cells in the logical table:

	A		A′	
	B	B′	B	B′
C	a	b	c	d
C′	e	f	g	h .

Since we know that the numerical entries corresponding to a, b, . . ., g, h will be probabilities it follows that we know that a + b + . . . + g + h = 1. Any information we have about the occurrence of A, B, or C, or of compound events involving the three simple events, can be expressed as an equation involving the eight letters in the cells of the table. Thus, we know that p(A) = .3 or a + b + e + f = .3, and that p(B) = .4 or a + e + c + g = .4, and that p(C) = .5 or a + b + c + d = .5, and p(AB) = .18 or a + e = .18, and that p(BC) = .28 or a + c = .28. These five equations plus the one that states that a + b + . . . + g + h = 1 give us six equations involving eight unknowns. Hence, we know that we can write our knowledge about the likelihoods of occurrence of the eight unknowns in the form of six equations involving two unknowns. I picked "b" and "e" as the unknowns and by straightforward solution and substitution I get the equations:

$$
\begin{array}{rcl}
a = u - e & \text{or} & a = .18 - e \\
c = v - u + e & \text{or} & c = .10 + e \\
d = C - v - b & \text{or} & d = .22 - b \\
f = A - u - b & \text{or} & f = .12 - b \\
g = B - v - e & \text{or} & g = .12 - e \\
h = 1 + u + v - A - B - C + b & \text{or} & h = .26 + b\,.
\end{array}
$$

Notice that one has a simple check on the correctness of the equations at this point:

$$a + b + c + d + e + f + g + h = 1$$

or, substituting

$$.18 - e + b + .10 + e + .22 - b + e + .12 - b + .12 - e + .26 + b = 1\,.$$

We follow Boole's procedure for compound events and put s = AB and u = p(s) = .18, and t = BC and v = p(t) = .28. We now have five simple events and we could present the logical table as having 25 = 32 cells but all except 8 of the cells would be 0 because they would be logically impossible. This follows because s = AB and t = BC are defined in terms of the original eight simple events. We do not have to define a specific variable in which we are interested because it is our intention to reconstruct the entire logical table. Once we have that then we can calculate the probability of any compound event which is defined in terms of the three original simple events, A, B, and C. We want, as we did when following Boole's own procedure, the events that are logically possible, given that there are now five simple events: s, t, A, B, and C. We can obtain these events by extremely simple logical multiplications instead of the laborious expansion of w which we used above. It is easiest to do the multiplications in two steps. First, we know that s = AB, hence s′ = A′ + B′, and t = BC, hence t′ = B′ + C′. Therefore, we have immediately that st = ABC, s′t = A′BC, st′ = ABC′ and s′t′ = A′B′ + A′C′ + B′ + B′C′. Now we must complete our multiplications by multiplying by the eight combinations of A, B, and C and their negations. In so doing we will remember that each such combination is represented by exactly one cell of our logical table. Thus, multiplying the above four preliminary multiplications by ABC we find that only stABC = a is possible. Similarly, multiplying each of the four by A′BC shows that only s′tA′BC = c is logically possible. (Remember that any term involving a variable and its negation is logically impossible.) Continuing these multiplications we find:

stABC	= ABC	= a
st′A′BC	= A′BC	= c
s′tABC′	= ABC′	= e
s′t′A′B′C	= A′B′C	= d
s′t′A′B′C′	= A′B′C′	= h
s′t′AB′C	= AB′C	= b
s′t′AB′C′	= AB′C′	= f
s′t′A′BC′	= A′BC′	= g .

The identical argument that Boole constructed applies to our list of eight logically possible combinations of our five simple events. Notice that our list of eight is exactly the same as we found when we expanded w above. The corresponding equations follow from Boole's own argument:

$$\frac{stABC}{a} = \frac{st'ABC'}{e} = \frac{s'tA'BC}{c} = \frac{s't'A'B'C}{d} = \frac{s't'A'B'C'}{h} =$$
$$= \frac{s't'AB'C}{b} = \frac{s't'AB'C'}{f} = \frac{s't'A'BC'}{g} = V$$

Where Boole's procedure produced five similar fractions equal to V, our approach gives eight fractions equal to V. Having a greater number of such fractions always makes it easier to solve the equations. It also makes it much easier to solve for the values of the Boolean variables.

We generally do not need to find the values of the Boolean variables for our emended method. However, we do need to find these values in order to complete the solution according to Boole's own method. How do we determine them? Find any pair of fractions for which the numerators have only one variable in one numerator and one minus the variable in the other numerator. For example, the fourth fraction's numerator contains C and the fifth fraction's numerator contains C′ = 1 - C. Multiplying both fractions by d and dividing both fractions by the numerator of #5 gives C/C′ = d/h or C = d/(d + h) and C′ = h/(d + h). Using the equations for d and h we find that d + h = .48 so C = d/.48 and C′ = h/.48. Similarly, from #5 and #7 we find A = f/(f + h) = f/.38 and A′ = h/.38. From #5 and #8 we calculate that B = g/(g + h) and B′ = h/(g + h). There are not two fractions with numerators only differing by one having s or t and the other having s′ or t′. Therefore, we will work with numerators differing by two (or more) variables. Thus, from #1 and #2 we can obtain tC/t′C′ = a/e. Hence, substituting for C/C′ we find that t/t′ = af/eb so t = af/(af + eb) and t′ = eb/(af + eb). Similarly, from #1 and #3 we find sA/s′A′ = a/c and substituting for A/A′ we find s/s′ = ah/cf so s = ah/(ah + cf) and s′ = cf/(ah + cf).

Having these values for the Boolean variables is by no means equivalent to having solved the equations. Here the difference between Boole's typical desideratum and ours becomes important. Suppose that for our example Boole wants, as he often does, only x = p(w) = stABC/V. Then we now know the values of the five Boolean variables and we can write

$$\frac{\frac{ah}{ah+cf} * \frac{ch}{ch+dg} * \frac{f}{f+h} * \frac{g}{g+h} * \frac{d}{d+h}}{x} = V$$

but this is not immediately useful. If we write it out in terms of the equations for a, . . ., h it becomes quite nightmarish: nothing cancels and nothing simplifies significantly. Nonetheless, the equation does suggest possibilities. Suppose we had been given x = p(w) = p(ABC) as part of our input information. Then, if we can evaluate stABC we will immediately know V, which can be used in any desired calculation. Otherwise, if we can evaluate stABC and know V then we can immediately find x, which quite often is the question which Boole wants to answer. We note several points. First, stABC occurs in each of the fractions in the Boolean equation and two of the fractions have only two terms each in the numerators so evaluation may not be as hopeless as it initially appears. Second, in our equations for a, . . ., h we have two unknowns, b and e. If we only had one unknown then the evaluation of stABC and other Boolean terms would be easier. Third, simplifications are often possible. Thus, in the first fraction (with p as denominator) we have s′t′AB′C + s′t′AB′C′ and these two terms = s′t′AB′. Boole's Example V, discussed as the penultimate example in this Chapter is a good example of the possibilities for simplification. Fourth, if we could solve for the two unknowns, b and e, then things would proceed much more smoothly.

We emphasized repeatedly that it is quite difficult to solve equations like these and our example is typical in this regard. Further, we commented that Boole is not at all helpful. He never says how he solved the equations. As an example, see (LT, p. 285) where Boole states the equations for an example that is roughly of the same level of difficulty as our example. Then he writes, "The solution of this system of equations gives" and gives the answer. Apparently, Boole thought that this was "just algebra" and that it would be demeaning to suggest that his readers could not solve such equations on sight. In any case, there is no question that such equations are algebraically complicated. We do well to avoid them wherever possible. Thanks to our emended method, we can always avoid them. The fourth comment above is precisely the easiest approach and this is exactly what the emended method does.

We now return to the emended procedure. We have eight fractions with denominators a, b, . . ., h. All of these fractions are equal to each other and each of them is equal to V. The fundamental difference in our present approach is that we intend to reason in terms of the denominators of these fractions. In order to accomplish our purposes we want to find two or more of the fractions which, when multiplied together, have the same resultant numerator as the numerator of another fraction or of the product of two or more other fractions. It is ordinarily not difficult to find such numerators and the simplest procedure is merely to contemplate the possibilities. Since we have two unknowns in our equations we will need to find two such relationships. In the present case I notice that the product of the 4th and the 7th of the fractions has numerator $(s')^2(t')^2AA'(B')^2CC'$ and that the product of the 5th and 6th of the fractions has exactly the same numerator. This means that

$$4*7 = \frac{(s')^2(t')^2AA'(B')^2CC'}{df} = 5*6 = \frac{(s')^2(t')^2AA'(B')^2CC'}{bh}$$

Hence, we know that df = bh. Replacing these by their equations we get (.22 - b)(.12 - b) = b(.26 + b) or .6b = .0264 and b = .044.

In finding a second relationship it is often very helpful to determine what the first relationship is in terms of the three simple events. We have d = A'B'C with f = AB'C' and b = AB'C with h = A'B'C'. In other words, within B' we have the well-known relationship A'C*AC' = AC*A'C'. This relationship is always true of two random and independent variables so the relationship that we have discovered says that within B' the two simple events A and C function as random and independent variables. This immediately suggests that it is worth checking whether A and C function similarly within B. If they do then we would expect that the product of A'BC = c and ABC' = e ought to equal the product of A'BC' = g and ABC = a. We quickly find that 2*3 does = 1*8:

$$2*3 = \frac{ss'tt'AA'B^2CC'}{ce} = 1*8 = \frac{ss'tt'AA'B^2CC'}{ag}$$

Therefore, ce = ag or e(.10 + e) = (.18 - e)(.12 - e), whence .4e = .0216 and e = .054. Now we have found the numerical values of each of the two unknowns. Therefore, we can calculate every entry in the logical table directly from our equations for a, c, d, f, g, and h:

	A		A′	
	B	B′	B	B′
C	.126	.044	.154	.176
C′	.054	.076	.066	.304 .

It is easy to verify that all of the input data are exactly recaptured. The emended method has made simple the standard kind of analysis that Boole's method made dismayingly difficult. This statement is by no means intended to derogate George Boole. On the contrary, the emended method is Boole's method and I have not the slightest doubt that this kind of improvement would have been found many years ago if there had been a "normal" development of the field that George Boole founded.

We will now use some of Boole's examples as illustrations of the emended method presented in this Chapter.

Section 20. Some of Boole's Examples, Based on Two Events

We have already given an example, based on mail-order merchandise premiums, that is exactly analogous to Boole's example of white and marble balls drawn from an urn. Therefore, we will not discuss this example further here. However, we will have occasion to consider it again in Chapter IV.

Every probability problem involving two random events has the same basic logical table. Assuming that the two events are X and Y, we can write the table thus:

	X	X′
Y	a	b
Y′	c	d

Boole's Example 1 of this kind of problem (LT, pp. 276-7) states that the probability that it thunders on a particular day is p and the probability that it both thunders and hails is q. Required is the probability that it hails on the proposed day. We are given, then, that P(X) = p and P(XY) = q. We have

$a + c = p$
$a = q$
$a + b + c + d = 1$

We want P(Y) = a + b.

We can solve for three of the variables:

$a = q$
$c = p - q$
$b + d = 1 - p$ or $b = 1 - p - d$.

Clearly, we cannot solve for b separately. Introducing a variable, k, with $0 \leq k \leq 1$ we can conclude that $b = k(1 - p)$. Hence, we can write $p(Y) = q + k(1 - p)$, as Boole does. This is an example in which the problem can be easily solved directly from the logical table and Boole's method is not needed.

Another such example is Boole's #2 (LT, p. 278). This example is given in Appendix B and inspection of that document is sufficient to show that a very great amount of algebra is required to apply Boole's method to this problem. On the other hand, the solution is immediate directly from the logical table. The problem is stated thus: The probability that one or both of two events happen is p and the probability that one or both of the two events fail is q. What is the probability that only one of the events happens. In terms of the logical table at the beginning of this Section, we know

$a + b + c = p$

$b + c + d = q$

$a + b + c + d = 1$

and we want b + c. But the first of these equations plus the second equation and minus the third equation gives $b + c = p + q - 1$, as Boole found after considerable labor.

Boole's example #3 (LT, pp. 279-281) involves the combination of testimony of two witnesses. The problem is stated thus: the probability that a witness A speaks the truth is p, the probability that another witness B speaks the truth is q, and the probability that they disagree in a statement is r. What is the probability that if they agree, their statement is true? Keynes maintains that all of the solutions given for this kind of problem are wrong, including Boole's, since Keynes argues that it is fallacious to assume that the joint probability that two independent witnesses both speak the truth equals the product of the separate probabilities that each of them speaks the truth.[14] However, it needs to be borne in mind that at the time Boole was writing and for the first half of the 19th century this kind of problem was quite widely discussed. Therefore, Boole was merely following well-established precedent. If we ignore this criticism then it is easy to obtain Boole's solution. The information we are given is (letting X be witness A and Y be witness B)

$$a + b = p$$
$$a + c = q$$
$$b + c = r$$
$$a + b + c + d = 1$$

Since we have four equations and four unknowns we can solve for all of the unknowns. We find:

$$a = (p + q - r)/2;\ b = (p + r - q)/2;\ c = (q + r - p)/2;\ d = (2 - p - q - r)/2\ .$$

The probability that the two witnesses agree is $a + d = 1 - r$ from the 3rd and 4th equations or directly from the values of the individual variables. The probability that they are both telling the truth is $a = (p + q - r)/2$ so the conditional probability that they are both telling the truth given that they agree is $a/(a + d) = (p + q - r)/2(1 - r)$. If you look at Boole's solution you will see that a considerable amount of algebra was required. Further, Boole then asks for the conditional probability that if the two witnesses agree the statement is false. Since he does not reason in terms of the underlying logical table, he has to make the whole analysis again. We can, of course, answer such an additional question immediately: the desired conditional probability is $d/(a + d) = (2 - p - q - r)/2(1 - r)$.

We already cited this example earlier since there is no doubt that Boole erroneously believed that it could not be solved by "known methods". In fact, each of the two-event examples cited in this Section can be solved by orthodox probability theory - the "known methods" to which Boole is referring. Whenever this is true then there is no need for Boole's method.

Section 21. Some of Boole's Examples, Based on Three Events

Chapter III will be devoted to the Challenge Problem and this, as well as its variations, is based on three events. In the present Section, then, we intend to look at a few of Boole's examples which involve three events and which are significantly different from the Challenge Problem.

The basic logical table that we will use for the examples in this Section is

	A		A′	
	B	B′	B	B′
C	a	b	c	d
C′	e	f	g	h

We may begin by looking at Boole's example 4 (LT, pp. 281-4). The simple events are A = house visited by fever; B = house visited by cholera; C = house in defective sanitary condition. The known probabilities are $P(A) = p$, $P(B) = q$, and $P(A'B'C') = r$. The desired probability is $P(C)$. The equations expressing the known information are:

$$a + b + c + d + e + f + g + h = 1$$

$$a + b + e + f = p$$

$$a + c + e + g = q$$

$$h = r$$

The solution of these equations can be written:

$$h = r$$

$$a + e = p - b - f$$

$$c + g = q - p + b + f$$

$$d = 1 - r - q - b - f \quad .$$

From the equations it is immediately clear that the unknown that can be found is b + f. “b” and “f” cannot be separately determined, granted the available data, nor can “a” and “e”, nor “c” and “g”. Since the desired p(C) is

$$p(C) = a + b + c + d$$

$$= 1 - r - e - f$$

it is obvious that it cannot be exactly determined. Following Step 6, we will put t = A’B’C’ and the three events with known probabilities are A, B, and t. There are, therefore, eight possible compound events such as tAB, tAB’, and so forth. Proceeding to the logical multiplication we find

$$tA'B' = A'B'C' = h$$

$$t'AB = ABC + ABC' = a + e$$

$$t'AB' = AB'C + AB'C' = b + f$$

$$t'A'B = A'BC + A'BC' = c + g$$

$$t'A'B' = A'B'C = d\ .$$

The other three combinations of t, A, and B and their negations are logically impossible. Thus, V = sum of the five terms on the left-hand sides of the above equations. The Boolean equations are:

$$\frac{tA'B'}{h} = \frac{t'AB}{a + e} = \frac{t'AB'}{b + f} = \frac{t'A'B}{c + g} = \frac{t'A'B'}{d} = V$$

From our linear equations we know that all we need to find is b + f. We are not at all interested in the numerical values of t, A, and B. By inspection we find that if we multiply the 2nd and 5th of these expressions and equate this to the product of the 3rd and 4th of the expressions, we have

$$\frac{(t')^2AA'BB'}{d(a+e)} = \frac{(t')^2AA'BB'}{(b+f)(c+g)}$$

Therefore, d(a + e) = (b + f)(c + g) .

Substituting from our linear equations in terms of (b + f), we get

(b + f)(q - p + (b + f)) = (1 - r - q - (b + f))(p - (b + f))

This is a linear equation in (b + f) and the solution is:

b + f = p(1 - r - q)/(1- r)

whence d = (1 - r - q)(1 - r- p)/(1 - r)

a + e = pq/(1 - r)

c + g = q(1 - p - r)/(1 - r)

and, of course, h = r.

The desired answer can be written in a variety of ways. The simplest is that since p(C) = 1 - r - e - f, it follows that p(C) = (1 - r) minus an undetermined proportion of (a + e) minus an undetermined proportion of (b + f). Boole's own solution is achieved by noting that since p(C) = a + b + c + d, it follows that p(C) equals "d" plus an undetermined proportion of p(A) (= a + b + e + f) plus an undetermined proportion of p(A'B) (=c + g). Of course, it can equally well be stated that p(C) equals "d" plus an undetermined proportion of p(B) (= a + c + e + g) plus an undetermined proportion of p(AB') (= b + f). Boole chooses the former and, therefore, writes

p(C) = ((1 - p - r)(1 - q - r)/(1 - r)) + k1p + k2q(1 - r - p)/(1 - r) .

This is his first expression for p(C). He then gives another expression by amalgamating the two terms multiplied by k's.

Obviously, the equation is exactly right in terms of Boole's method. Hailperin, in his discussion of this example states: "The application of standard techniques does not reproduce Boole's result: . . ." He goes on to say: "It appears that Boole's result is in error. But not badly so!"[15] This is particularly interesting because this is one of two of the elementary examples that happens to have a probability that cannot be calculated by standard procedures but can be calculated by Boole's method. In other words, this particular example shares with the Challenge Problem the fact that Boole is claiming to be able to calculate a probability that cannot be calculated by the usual theory. Exactly as Boole's contemporaries criticized his solution of the Challenge Problem as being in error, so is Hailperin criticizing Boole's solution of this problem as being in error. Always, the implication is that Boole is wrong because his solution does not correspond with orthodox probability theory. But Boole said as clearly as it is possible to be said that orthodox probability theory was not adequate and could not provide answers to many questions that Boole maintained were reasonable questions. He developed his system precisely to answer these questions and it does answer them. Therefore, it was and remains totally beside the point to criticize Boole because his solutions do not correspond to orthodox solutions. He would not have succeeded in his efforts if they did. A critic can certainly argue that Boole's whole system is wrong-headed and erroneous. But if so, reasons must be given that attack the assumptions that are the basis of his system, not solutions that are exactly correct, granted his system.

How did Hailperin reach his conclusion that Boole's solution to the present example is wrong? He uses ordinary probability theory to deduce an expression for p(C).[16] Having rewritten Boole's expression for p(C) in an analogous form, he correctly points out that the two expressions differ only in that, where Boole's expression has pq/(1 - r), Hailperin's expression has p(xy). The only thing wrong with this statement is the word "only" (my word, not Hailperin's). The essence of Boole's whole point lies in this difference. Put as simply as possible, p(xy) is not given as part of the data so Hailperin's "solution" is not a solution in Boole's sense at all. They could not be more different.

I must emphasize that this difference is always present when a comparison is being made between Boole's theory and the usual theory. Specifically, if there are unknowns on the right-hand sides of the linear equations based on a, b, . . . , h then the problem cannot be solved by orthodox theory. This is without any exceptions. Boole's method will, of course, generally produce a solution in such cases.

I will not comment upon Boole's examples 5 and 6 because I believe that Hailperin's discussion of these examples is very much to the point and there is nothing I can add to it.[17]

All of the examples used in Section 20 plus the preceding example are from Boole's Chapter XVIII, "Elementary Illustrations of the General Method in Probabilities". One final example from the same Chapter is worthy of discussion. This is the third case of Example 7 (LT, pp. 290-4). This is the hardest of the examples of Chapter XVIII and Boole's own analysis is complicated and lengthy. The great superiority of the procedure of Section 18 will be well shown by this example.

We are given p(A + A'BC) = p, p(B + AB'C) = q, and p(C+ ABC') = r. This is another instance where Boole has deliberately selected difficult and unrealistic given probabilities with the intention, no doubt, of thus demonstrating the power of his method. The equations for the known information are:

$$a + b + e + f + c = p$$

$$a + c + e + g + b = q$$

$$a + b + c + d + e = r$$

$$a + b + \ldots + g + h = 1$$

The desideratum is p(A).

The solution can be written:

$$d = r - a - b - c - e$$

$$f = p - a - b - c - e$$

$$g = q - a - b - c - e$$

$$h = 1 - p - q - r + 2(a + b + c + e)$$

The unknown is (a + b + c + e) and it is clear that we cannot separately determine these terms with the given information. From this it immediately follows that we will not be able to exactly determine p(A).

Putting s = A + A′BC, t = B + AB′C, and u = C + ABC′ it follows that s′ = A′B′ + A′C′, t′ = A′B′ + B′C′, and u′ = A′C′ + B′C′. From these, by logical multiplication we find:

stu = ABC + A′BC + AB′C +ABC′ = a + b + c + e

s′t′u = A′B′C = d

st′u′ = AB′C′ = f

s′tu′ = A′BC′ = g

s′t′u′ = A′B′C′ = h

V is the sum of the five terms on the left-hand side of the equations. The other combinations of s, t, and u and their negations are logically impossible.

The Boolean equations are:

$$\frac{stu}{a+b+c+e} = \frac{s't'u}{d} = \frac{st'u'}{f} = \frac{s'tu'}{g} = \frac{s't'u'}{h} = V$$

Here all we need to find is (a + b + c + e). By equating the product of the 5th expression squared and the 1st expression to the product of the 2nd, 3rd, and 4th, we get

$$\frac{stu(s't'u')^2}{h^2(a+b+c+e)} = \frac{stu(s't'u')^2}{dfg}$$

Whence, $h^2(a + b + c + e) = dfg$.

Substituting for h, d, f, and g in terms of (a + b + c + e) gives, putting z = a + b + c + e temporarily

$$z(2z + 1 - p - q - r)^2 = (p - z)(q - z)(r - z)$$

The solution of this cubic equation gives the complete solution of the logical table to the degree that it can be determined from the given information.

Boole's solution is also a cubic equation but in an intervening variable (λ). It is impossible to algebraically demonstrate the identity of the two solutions so perhaps a numerical example would be helpful. Suppose p = .6, q = .5, and r = .4. The solution of the cubic equation given above is

d = .0881227

f = .2881227

g = .1881227

$$h = .1237546\ .$$

The desired probability is indeterminate but we can say that $P(A) = f + kz$, or

$$p(A) = .2881227 + k(.3118773)\ .$$

Boole's solution has $\lambda = 8.08050767$ with exactly the same values in the logical table as are given above. If the reader will take the time to look up Boole's own analysis I am sure he or she will immediately see the very great simplification of the procedure given in Section 18.

In his Chapter XIX, "Of Statistical Conditions", Boole considers the question he states as: "By the term statistical conditions, I mean those conditions which must connect the numerical data of a problem in order that those data may be consistent with each other, and therefore such as statistical observations might actually have furnished." (LT, p. 295) This is a major question and Boole's answers are, as he knew, incomplete. However, it is surely true that Hailperin has answered the general question with neatness and dispatch.[18] Therefore, it is not necessary to discuss this question here. This is particularly true since my interest is solely in the applications of Boole's method to real-world problems. In any such problem it is a matter of definition that the "conditions of possible experience" are satisfied.

Boole's Chapter XX, "Problems on Causes" gives a number of additional examples of his method. It is among these examples that we find the Challenge Problem and variations on it. As the title of the Chapter shows, Boole puts these examples together because he believes that they all relate in one way or another to "the study of the relation of cause and effect in the more complex of its actual and visible manifestations." (LT, p. 321) It is not at all clear whether the distinction he has in mind is an important one. Boole, along with most of his contemporaries, believed in "the pre-eminence which is due, above all other relations, to the relation of cause and effect" (LT, p. 320). But today we are perhaps more restrained with regard to this relation. The whole notion is philosophically opaque. Virtually the only relevant proposition on which there is general agreement is that the cause must temporally precede the effect and even concerning this there are serious doubts in modern particle physics. Further, when we lean heavily on the temporal precedence aspect of the relation we often fall into the fallacy of *post hoc ergo propter hoc*. Because of all the difficulties associated with the relation, very few serious works on science or philosophy waste much space talking about cause and effect.

The point of these remarks is that, regardless of what Boole may have thought, the examples of his Chapter XX, when stripped of the verbiage of the cause-and-effect relation, become ordinary problems exactly similar to those of Chapter XVIII, already discussed. Of course, there are some differences. Most of these examples are somewhat more complicated than the elementary examples of Chapter XVIII. Correspondingly, the deduction of "conditions of possible experience" becomes more difficult and perhaps more important. However, with regard to the key point of the difference between orthodox probability theory and Boole's method, there is no difference between the examples of Chapter XX and those of Chapter XVIII. There are at least two elementary examples that cannot be handled by the usual theory and for which Boole's method provides exact estimates. This is generally the case and the examples of Chapter XX should not be considered exceptional in this regard.

The next Chapter of this book is devoted to the Challenge Problem and variations of it so we do not need to consider them in the present Section. Here we will discuss two of his examples in Chapter XX (LT, pp. 334-6). We could also discuss some analysis he made elsewhere of his example of rain, snow, and wind, previously mentioned in the context of his explanation of the reason compound events should be treated similarly to simple events. Instead, we will postpone that discussion to Chapter IV.

In Boole's problem IV of Chapter XX we are given p(A) = p, p(B) = q, p(C) = r, p(AC) = s, and p(BC) = t. Required is p(AB). It may be noted that this is exactly the Challenge Problem with two changes. First, here we are given p(C) whereas in the Challenge Problem this is the desideratum. Second, here we are asked for p(AB). The equations expressing our knowledge are:

$$a + b + e + f = p$$

$$a + c + e + g = q$$

$$a + b + c + d = r$$

$$a + b = s$$

$$a + c = t$$

$$a + b + c + d + e + f + g + h = 1 .$$

The solutions of these equations can be written:

$$b = s - a$$

$$c = t - a$$

$$d = r - s - t + a$$

$$f = p - s - e$$

$$g = q - t - e$$

$$h = 1 - p - q - r + s + t + e$$

Since there are unknowns on the right-hand sides, we know that there is no way to achieve a unique estimate by the usual probability theory.

Set u = AC and V = BC. Then u' = A' + C' and v' = B' + C'. By ordinary logical multiplication we have:

$uvABC$	$= ABC$	$= a$
$uv'AB'C$	$= AB'C$	$= b$
$u'vA'BC$	$= A'BC$	$= c$
$u'v'ABC'$	$= ABC'$	$= e$
$u'v'AB'C'$	$= AB'C'$	$= f$
$u'v'A'BC'$	$= A'BC'$	$= g$
$u'v'A'B'C$	$= A'B'C$	$= d$
$u'v'A'B'C'$	$= A'B'C'$	$= h$

Of the 32 possible combinations of u, v, A, B, and C and their negations, only the eight given above are logically possible. The sum of these eight, then, is V. The Boolean equations are:

$$\frac{uvABC}{a} = \frac{uv'AB'C}{b} = \frac{u'vA'BC}{c} = \frac{u'v'ABC'}{e} =$$
$$\frac{u'v'AB'C'}{f} = \frac{u'v'A'BC'}{g} = \frac{u'v'A'B'C}{d} = \frac{u'v'A'B'C'}{h} = V$$

These equations look ominous but they are really quite easy to solve. Remember that we need only "a" and "e". We can completely ignore the Boolean variables u, v, A, B, and C. A reasonably quick procedure for solving equations of this type is to simplify the equations by dividing each of the numerators by uvABC and solving for the resulting unknowns of the form A'/A. Here it is sufficient to note that equating the product of the 2nd and 3rd of these expressions to product of the 1st and 7th gives

$$\frac{uu'vv'AA'BB'C}{ad} = \frac{uu'vv'AA'BB'C}{bc}$$

so $$ad = bc$$

or $$a(r - s - t + a) = (s - a)(t - a)$$

whence $$a = st/r\ .$$

Similarly, equating the product of the 5th and 6th expressions to that of the 4th and 8th gives

$$eh = fg$$

or $$(p - s - e)(q - t - e) = e(1 - p - q - r + s + t + e)$$

whence $$e = (p - s)(q - t)/(1 - r)\ .$$

The problem is solved since all of the entries in the logical table are now known. Since the desired probability is p(AB) = a + e the answer is simply

$$p(AB) = (st/r) + (p - s)(q - t)/(1 - r)\ .$$

This is, of course, identical to Boole's solution.

Problem #5 is given very brief discussion by Boole. I suspect that he included it only because it is very much the hardest of any of his examples that involve three or less simple events. Boole gives almost no information as to how he achieved his solution but the mere form of it is sufficient to suggest that he probably had to work long and hard to get it. In any case, the given probabilities are p(A) = p, p(B) = q, p(C) = r, and p(ABC) = m. Required is the probability of occurrence of any one of the events: AB, AC, or BC.

The difficulty of the solution has only to do with solving the final equations. Everything else proceeds as easily as usual. Thus, the solutions of the logical equations can be written

b = p - m - e - f

c = q - m - e - g

$$d = r - p - q + m + 2e + f + g$$

$$h = 1 - r - e - f - g$$

And, of course, a = m. Here there are three unknowns: e, f, and g.

Putting u = ABC, whence u′ = A′ + B′ + C′, and performing the logical multiplications gives:

$$\begin{aligned}
&uABC &&= ABC &&= a\\
&u'ABC' &&= ABC' &&= e\\
&u'AB'C &&= AB'C &&= b\\
&u'AB'C' &&= AB'C' &&= f\\
&u'A'BC &&= A'BC &&= c\\
&u'A'BC' &&= A'BC' &&= g\\
&u'A'B'C &&= A'B'C &&= d\\
&u'A'B'C' &&= A'B'C' &&= h\,.
\end{aligned}$$

The sum of the terms on the left-hand sides of the equations is V. The Boolean equations are written as usual and will not be given here.

It is quite easy here to deduce equations among a, b, . . . , h and, hence, equations for the unknowns, e, f, and g. This, of course, means that it would generally be straightforward to use iterative, or other numerical, methods to resolve any such actual problem. However, if we stick to the rules that Boole was forced to observe because of the lack of computational assistance then we, too, would like to find an explicit formula for the solution. The equations that can be deduced are:

$$de = fc = bg;\ fg = eh;\ fd = bh;\ dg = ch$$

The first set of three equalities has an interesting fourth member, hbe/f. This means that the Boolean equation for hbe divided by f is the same as the Boolean equations for the other three. Setting de = beh/f, dividing through by "e", and multiplying through by "f" gives an equation which leads directly to the equation for "e" immediately below. The solution for "e" is that "e" is the root of the quadratic:

$$(-1 + r)E^2 + ((p - r)(q - r) + (1 - r)(1 + r - 2m))e - (1 - r)(p - m)(q - m) = 0\,.$$

This is the same as the equation that Boole deduces, recognizing that our equation is for "e" while Boole's equation is for (a + e).

Then, $g = e(1 - p - q + m + e)/(p - m - e)$

and $f = e(1 - p - q + m + e)/(q - m - e).$

This is noticeably simpler than Boole's solution and, of course, is numerically identical to his. (See the footnote for some comments on obtaining these equations.)[19] As a numerical example, suppose p = .4, q = .3, r = .2, and m = .05. The complete solution is:

	A		A′	
	B	B′	B	B′
C	.05	.046067	.028745	.075188
C′	.084059	.219875	.137197	.358870

Perhaps it is worth stating again: the usual probability theory cannot accomplish any of these estimates. Boole's method, as we see, can. There is no way whatsoever that anyone can make a valid and meaningful comparison between Boole's theory and the orthodox theory because the latter cannot do the things that are done in the former. This is why that all of the criticisms of Boole's theory from Macfarlane and MacColl through Keynes and on to, but not including, Hailperin are not to the point. Boole may be wrong but if so it needs to be demonstrated, not just talked about. Note that this statement includes the argument of Wilbraham and Cayley, which will be discussed at length in Chapter IV.

Boole provides hardly any information as to how he solved his Problem V. This is somewhat unfortunate since the format of his answer is quite different in appearance from any of his other solutions. I have never been able to reconstruct his reasoning. His whole discussion, including his answer is somewhat less than half a page (see LT, p. 335).

Earlier, it was mentioned that the example just presented provided a good illustration of the possibilities of finding V arithmetically by simplifying the summation required. As we calculated above, there are eight terms, from uABC through u'A'B'C', which need to be summed in order to find V. However, we can use the elementary properties of logical variables to lump them as follows:

7 & 8 : u'A'B'C + u'A'B'C' = u'A'B'

5 & 6 : u'A'BC + u'A'BC' = u'A'B

So, 5 & 6 & 7 & 8 : = u'A'. Similarly, 3 & 4 = u'AB'.

Instead of having to calculate the numerical values of eight quantities all we need is the sum of 1 & 2 & 3-4 & 5-8. Using the known numerical quantities given for this example we quickly find that u = .739518, A = .379917, B = .276570, and C = .173222. These values give uABC = .0134600, u'ABC' = .0226288, u'AB' = .0715918, and u'A' = .1615204. The total is V = .2692010. Note that we are given that p(ABC) = .05 so we have also that uABC/.05 =.269200 = V. We can then calculate each entry in the logical table directly. Thus, u'ABC'/.269200 = .084059 = e. Obviously, this kind of calculation is only worthwhile if a specific example needs analysis. If several having the same structure need analysis then it is ordinarily much easier to find a general result which can be repeatedly applied.

After his very brief discussion of problem #5, Boole adds the comment: "Had the condition, P(C) = r, been omitted, the solution would still have been definite." (LT, p. 335) (Notation changed to conform to that used here.) He then gives the solution for p(AB). The contrast in the solutions of the two cases is interesting and will be useful subsequently so I will briefly indicate the solution of this version of Problem 5. The solution of the linear equations is:

a = m

$e = p - m - (c + g)$

$b + f = q - p + (c + g)$

$d + h = 1 - q - (c + g)$

The unknown is (c + g) and it is clear that "c" and "g" cannot be separately determined with the information available, nor can "b" and "f", nor "d" and "h". Nonetheless, since p(AB) = a + e, this probability can be exactly estimated. Presumably, it was this point that Boole desired to make with the present example. Proceeding directly to the Boolean equations, we find, with u = ABC:

$$\frac{uAB}{m} = \frac{u'AB}{e} = \frac{u'AB'}{b+f} = \frac{u'A'B}{c+g} = \frac{u'A'B'}{d+h} = V$$

Note that C does not appear as a Boolean variable because we do not know P(C). Writing the product of the 2nd and 5th expressions equals the product of the 3rd and 4th and equating the denominators gives:

$$e(d + h) = (b + f)(c + g)$$

or $$(c + g)(q - p + c + g) = (p - m - c - g)(1 - q - c - g) \quad .$$

This is a linear equation in (c + g) and the solution is:

$$c + g = (p - m)(1 - q)/(1 - m) \; .$$

Therefore,

$$e = (p - m)(q - m)/(1 - m)$$

and

$$p(AB) = \frac{pq + (1-p-q)m}{1-m}$$

For purposes of comparison, the previous example (Boole's original Problem #5) gave p(AB) = .134059. The present solution, assuming that we do not know p(C) = .2, gives p(AB) = .142105.

We will subsequently discuss another example of Boole's which could be included here. However, this example is related to Boole's argument about speakers of a dialect which uses the words "sleet", "storm", and "drift" and which has no simple words for "snow", "rain", and "wind". Boole uses this example to make his point that what we consider to be "simple" events as contrasted with "compound" events may be more nearly a matter of our linguistic usage than it is a "real" difference. Therefore, we will more appropriately discuss this example in Chapter IV.

CHAPTER III

THE CHALLENGE PROBLEM

Section 22. The Challenge Problem.

The circumstances surrounding the Challenge Problem were discussed in Section 1 and those details will not be repeated here. Instead, we will consider, compare, and evaluate the various solutions that were suggested.

Let us state the Challenge Problem in a form appropriate for the logical tables we have been using. Since three variables are involved this is easy to do. We are given that the probability of A is p and the probability of B is q. Further, we are given that the joint probability of AC is r and the joint probability of BC is s. Finally, we are told that C cannot occur unless at least one of A and B occurs. The question, then, is: What is the probability of C?

The logical table is, as always:

	A		A'	
	B	B'	B	B'
C	a	b	c	d
C'	e	f	g	h

The information we have leads to the following equations:

$$a + b + c + d + e + f + g + h = 1$$
$$a + b + e + f = p$$
$$a + c + e + g = q$$
$$a + b = r$$
$$a + c = s$$

and

$$d = 0\ .$$

The last equation reflects the fact that A'B'C = 0 since C cannot occur unless at least one of A or B occurs. This is what Boole meant by conditioning. Variables are conditioned when it is impossible for some combination of them to occur or when it is necessary that some combination of them occur.

There are eight variables and only six equations. Therefore, it is obvious that we cannot obtain unique values for all of the variables. This was perfectly clear to Wilbraham and Cayley and to all of Boole's subsequent critics. Nonetheless, he claimed to be able to find unique values for all of the variables by his probability theory. Further, he claimed that the orthodox theory was inadequate precisely because it did not permit solving such problems.

It is certainly true, as Wilbraham emphasized, that the only way the equations can be solved is if two additional equations are added to the ones above. Cayley's solution had to have been based on some additional pair of equations and so did Boole's solution. Wilbraham implied that Boole was surreptitiously introducing his two additional equations and also that at least one of the equations so introduced was unreasonable. He further maintained that the two equations added by Cayley were reasonable. Boole indignantly denied that there was anything surreptitious about his procedure and he claimed that the two added equations were inherent in the data of the problem and directly and inevitably deduced by his logical method.

Subsequent critics shifted the grounds of the attack on Boole and his solution to the Challenge Problem. Instead of counting equations, with Wilbraham, they argued that something was wrong with Boole's definition of independence. This, as noted earlier, was Keynes's main argument against Boole. It is, in fact, a major lowering of the standards as compared to Wilbraham because there is no possible way that Boole's argument could depend on the incorrect definition of independence. Since this was discussed in Chapter I it does not need further discussion here.

There is no doubt whatsoever that Boole's main reason for proposing the Challenge Problem was precisely that his method gives a unique and specific answer for the probability of C while orthodox theory does not. As I remarked in Chapter I, Boole had "bad luck" in his choice of an example. The first part of the "bad luck" was that Cayley succeeded in finding a unique value for P(C) - presumably by using orthodox theory. We may begin our consideration of the Challenge Problem by looking at Cayley's solution to it.

Section 23. Cayley's Solution of the Challenge Problem.

Cayley's solution is nothing if not ingenious. He explicitly assumes that A and B are independent. Therefore, $p(AB) = pq$, $p(AB') = p(1 - q)$, and similarly for the other two possibilities. Now Cayley assumes that the probability that A acts "efficaciously" to produce C is λ and that the same probability for B is μ. Using the customary notation for conditional probabilities, Cayley writes:

$$p(C|A'B') = 0$$

$$p(C|AB') = \lambda$$

$$p(C|A'B) = \mu$$

$$p(C|AB) = 1 - (1 - \lambda)(1 - \mu) = \lambda + \mu - \lambda\mu .$$

Therefore, the joint probabilities are:

$$p(C,AB') = \lambda p(1 - q)$$

$$p(C,A'B) = \mu(1 - p)q$$

$$p(C,AB) = (\lambda + \mu - \lambda\mu)pq .$$

Immediately it follows that p(C) is the sum of these three probabilities, or

$$(1) \quad p(C) = \lambda p + \mu q - \lambda\mu pq .$$

Now, $p(C|A) = r/p$ on the one hand and on the other hand

$$p(C|A) = (p(C,AB') + p(C,AB))/p$$

$$= \lambda + (1 - \lambda)\mu q \quad .$$

Equating these two expressions for p(C|A) gives

$$(2) \quad r = p\lambda + p(1 - \lambda)\mu q \quad .$$

Working similarly with p(C|B) gives

$$(3) \quad s = q\mu + q(1 - \mu)\lambda p \quad .$$

The numbered equations are three equations in the three unknowns λ, μ, and p(C). Hence, they can be solved to find p(C). However, it is by no means obvious how to solve them. Here is Cayley's method. For convenience, put u = p(C).

(1) - (2) gives	$u - r = \mu(1 - p)q$	(4)
(1) - (3) gives	$u - s = \lambda p(1 - q)$	(5)
Subtract (4) from	$1 - p = 1 - p$	
and obtain	$1 - p + r - u = (1 - p)(1 - \mu q)$.	
Subtract (5) from	$1 - q = 1 - q$	
and obtain	$1 - q + s - u = (1 - q)(1 - \lambda p)$.	

Multiplying these last two equations together gives

$$(1 - p + r - u)(1 - q + s - u) = (1 - p)(1 - q)(1 - u)$$

since $(1 - \lambda p)(1 - \mu q) = 1 - u$ by (1). Thus, we have obtained a quadratic equation in u = p(C) and the desired probability has been found.

Here is a numerical illustration. Suppose p(A) = p = .5, p(B) = q = .6, p(AC) = r = .4, and p(BC) = s = .42. Then we find u = p(C) = .561003. Since we have not worked out our usual linear equations for the coefficients in the logical table it requires more effort to deduce the logical table. It is:

	A		A'	
	B	B'	B	B'
C	.258997	.141003	.161003	0
C'	.041003	.058997	.138997	.2

In cidentally, λ = .705013 and μ = .536675.

It may be noted now for future reference that it is very much easier to solve for Cayley's solution if we know the two equations that his method added to the six linear equations given in Section 22. As will be discussed subsequently, Cayley himself stated that the two added equations were

$$eh = fg$$

$$\frac{a + e}{q - e - a} = \frac{p - e - a}{1 - p - q + a + e}$$

Solving the six linear equations gives:

$$b = r - a$$

$$c = s - a$$

$$f = p - r - e$$

$$g = q - s - e$$

$$d = 0$$

$$h = 1 - p - q + a + e \quad .$$

These eight equations in eight unknowns are easy to solve. Solving the second of the equations added by Cayley for (a + e) gives

$$a + e = pq \; .$$

Substituting this in the first added equation gives a quadratic equation in “e”:

$$E^2 - e(1 + pq - r - s) + (p - r)(q - s) = 0 \quad .$$

Putting in the values from the preceding numerical illustration gives

$$e = .041003$$

whence

$$a = .258997$$

and all the other values follow. This procedure is clearly a considerable simplification of Cayley’s argument.

Section 24. Boole’s Solution of the Challenge Problem.

Boole's solution of the Challenge Problem results from a straightforward application of the procedure outlined in Section 18. We already have the solutions of the linear equations, with "a" and "e" as the unknowns. We introduce the two new variables t = AC and u = BC, whence t' = A' + C' and

u' = B' + C'. We introduce these variables because we know the probabilities of AC and BC. Since we do not know the probability of C, it is not a Boolean variable. Using t, u, A, and B and ordinary logical multiplication we find:

$$
\begin{aligned}
&tuAB &&= ABC &&= a\\
&tu'AB' &&= AB'C &&= b\\
&t'uA'B &&= A'BC &&= c\\
&t'u'AB &&= ABC' &&= e\\
&t'u'AB' &&= AB'C' &&= f\\
&t'u'A'B &&= A'BC' &&= g\\
&t'u'A'B' &&= A'B' &&= d + h
\end{aligned}
$$

Dividing the left-hand sides of these equations by the right-hand sides and equating all of them to each other and to V (the sum of the seven left-hand sides) gives the Boolean equations. The product of the 1st and the 7th of these has the same numerator as the product of the 2nd and 3rd. Hence,

$$bc = a(d + h) \quad \text{or} \quad (r - a)(s - a) = a(1 - p - q + a + e) \ .$$

We know that d = 0 so it could be dropped from the equation. I continue to include it for future reference. The product of the 4th and 7th of the Boolean expressions has the same numerator as the product of the 5th and 6th. Hence,

$$fg = e(d + h) \quad \text{or} \quad (p - r - e)(q - s - e) = e(1 - p - q + a + e) \ .$$

These two additional equations, added to the original six, give eight equations in eight unknowns.

Solving the first of the two additional equations for "a" gives:

$$a = \frac{rs}{1 - p - q + r + s + e}$$

Substituting this value of "a" in the second of the added equations and simplifying gives a quadratic for "e":

$$-(1 - r - s)E^2 + e((p - r)(q - s) - rs - (1 - r - s)(1 - p - q + r + s)) +$$

$$+ (p - r)(q - s)(1 - p - q + r + s) = 0 \ .$$

Using the values of the previous numerical illustration, we find

$$e = .045046$$

whence

$$a = .219595$$

and all of the other values follow. The reconstructed logical table is:

	A		A'	
	B	B'	B	B'
C	.219595	.180405	.200405	0
C'	.045045	.054954	.134954	.164640 .

Comparison of these values with those found from Cayley's solution shows that the two solutions are clearly different. The arithmetical magnitude of the difference is not large but, for example, Cayley's value for ABC is about 18% larger than Boole's. This is not a trivial difference.

In any case, it is clear that the solutions are different. Who, then, is right?

Section 25. Boole and Cayley: the Disagreement.

The disagreement between Boole and Cayley concerning the solution of the Challenge Problem began in early 1854. Cayley published his solution in an extremely abbreviated form in October, 1853.[1] Boole responded in January, 1854.[2] in this article Boole completed Cayley's solution and showed two examples for which Cayley's solution was undoubtedly correct. But, he proceeded to say: "I suppose that few persons, after reading Mr. Cayley's solution and noticing the above verifications, would feel any doubt of its correctness; and yet it is certainly erroneous."

As evidence of the incorrectness of Cayley's solution Boole proposes the case with $r = p$ and with $s = 0$. in words, this case implies that if A occurs then C certainly does but if B occurs then C never does. Boole claims that it is evident that this means $p(C) = p$ since C occurs if and only if A occurs. Boole notes that his own solution gives the correct answer. This can be easily verified from the solution as given in the preceding Section. It will be found that $e = 1 - q$, whence $a = 0$ and then $b = p$ and $c = 0$. Since $p(C) = a + b + c = p$ it is clear that Boole's solution does give the correct answer.

But Cayley's solution, as Boole points out, does not give this answer. Boole's own form of Cayley's solution is different than either of the two solutions given in Section 23 although, of course, equivalent to them. From his form he deduces that by Cayley's solution in the case under consideration it would follow that either $p(C) = 1$ or $p(C) = p(1 - q)$ and neither of these answers is correct. From the first of our two solutions follows that $p(C) = p(1 - q)$ and from the second of them follows that no solution is possible since negative probabilities result.

In all of the solutions it must be concluded that Cayley's solution does not give the correct answer but the differences among the solutions - each of which is otherwise correct - are an obvious indication that some basic mistake, like dividing by zero, must have been made. This is, in fact, true and Cayley himself pointed it out. in the somewhat curious chronology of the Cayley-Boole interaction on the Challenge Problem, there was no publication by Cayley after Boole's 1854 article until 1862 and this article has such extensive quotations from Boole that it might well be considered a jointly authored article.[3] in any case, in this article Cayley comments: "... In fact, the two equations $r = p$, $s = 0$ (i.e. if A then C always; if B, then C never) would imply that A and B could not happen together; which is contrary to the hypothesis that they are independent causes." (The notation has been changed to correspond to that used here.) This is certainly true and since Cayley explicitly assumes that A and B are independent this case constructed by Boole violates the assumptions of Cayley's solution. Boole apparently recognized this because he does not again advert to this counter-example.

In stead, in the article now under consideration, Boole claims that Cayley's solution is not in accord with the "conditions of possible experience" of the data of the Challenge Problem. Boole gave considerable attention to methods for determining the conditions which must be met by the known probabilities of various problems in order that there should be a possible solution. This part of Boole's work has been so ably completed by Hailperin that I have not considered it here.[4] For the Challenge Problem, however, the conditions are quite simple. Specifically, Boole states that it is necessary that

$$p - r + s \geq pq$$

and

$$q + r - s \geq pq$$

as well as the obvious conditions that each of p, q, r, and s should be probabilities. Boole goes on to state that Cayley's solution requires that

$$s \geq rq$$

and

$$r \geq ps \quad .$$

Boole says that these two sets of conditions are not equivalent and that Cayley's is more restrictive. Therefore, there are values of the known probabilities for which there are solutions but Cayley's argument will not find them. Further, Boole claims that his method will always produce a solution whenever the known probabilities could actually occur.

It is easy to see that Boole is correct on this issue. Suppose p = .6, q = .7, r = .4, and s = .25. It can be verified that these values satisfy Boole's "conditions of possible experience" but do not satisfy Cayley's requirements. Using the second solution given in Section 23 we find that

$$e = .1437$$

whence $a = .2763$

and $b = .1237$

but $c < 0 \quad .$

Therefore, Cayley's argument does not work with these values.

On the other hand, for these values Boole's solution gives

$$e = .165438$$

whence $a = .194010$

and the logical table can be reconstructed as:

	A		A'	
	B	B'	B	B'
C	.194010	.205990	.055990	0
C'	.165438	.034562	.284562	.059448 .

Therefore, Boole certainly has a solution for this case and Cayley does not. Boole apparently considers this to be a major argument against Cayley's solution.

Cayley, on the other hand, cheerfully admits that his solution requires more restrictive conditions than does Boole's. in fact, he says that for his argument: "the conditions of consistency ought to be narrower than those in Professor Boole's (solution)". The reason that Cayley so readily admits this fact that Boole had argued against him is that Cayley maintains that there are two versions of the Challenge Problem. The one that he says Boole stated and that he, Cayley, solved, Cayley calls the causation version. On the other hand, Cayley maintains that Boole solved a different version, that Cayley calls the concomitance version. Cayley recurs to this theme several times in the article under consideration. in the present context he argues that the more stringent conditions are both necessary and sufficient for the causation version and that Boole's conditions are only valid for the concomitance version.

Cayley ascribes great importance to this distinction so we need to try to understand it. His clearest statement of the difference seems to be this: "To show that the two statements are really distinct questions, it may be observed that when A and B both exist, then, according to the causation statement, they may one or each of them act efficiently, and C may thus happen as an effect of one of them only, or as an effect of each of them; but according to the concomitance statement, C cannot be attributed rather to one of the events A, B, than to the other of them, or to both of them." This quotation is not quite pellucid but its main thrust seems clear enough. His own solution very definitely depends on the causation version since his whole approach to the Challenge Problem is based on the assumption that C can be attributed to a specific one of the two possible causes. Therefore, among other things, Cayley argues that this immediately results in more stringent conditions on the known probabilities.

On the other hand, Cayley repeatedly states that he doesn't understand Boole's argument. in other words, he admits that he does not understand the concomitance version. Thus, Cayley states at the end of the article: "We thus join issue as follows: Professor Boole says that there is no difficulty in understanding, I say that I do not understand, the rationale of his solution."

How might Boole have responded to Cayley's argument? in the article under consideration, Cayley quotes three full pages of a letter from Boole to Cayley, dated March 26, 1862, so one might ask how Boole actually did respond. However, it seems to me that Boole did not state his case nearly as forcefully as he might have. Perhaps this was due merely to his natural courtesy and diffidence or, as I am inclined to suspect, it may have been due to the fact that he was dissatisfied with his argument. In any case, here I will state the salient points of Boole's actual response and then suggest some elaborations he could have given.

In the letter quoted by Cayley, Boole gives eight numbered statements. Of these the first and seventh are particularly germane to our purposes. His first statement is: "I think that your solution is correct under conditions partly expressed and partly implied. The one to which you direct attention is the assumed independence of the causes denoted by A and B. Now I am not sure that I can state precisely what the others are; but one at least appears to me to be the assumed independence of the events of which the probabilities according to your hypothesis are r and s. Assuming the independence of the causes as to happening, I do not think that you are entitled on that ground to assume their independence as to acting; because, to confine our observations to common experience, we often notice that states of things apparently independent as to their occurrence, may, when concurring, aid or hinder each other in such a manner that the one may be more or less likely to act `efficiently' in the presence of the other than in its absence. I use the language of your own hypothesis of efficient action."

In his seventh statement, Boole says: "The problem, as stated by me, and then modified by the simple introduction of the hypothesis of the independence of A and B, must admit of solution by my method; and that solution ought to impose no restriction beyond the conditions of possible experience ...".

With regard to his first statement, one must conclude that he is being diffident. At least after reading Wilbraham's article of 1854, Boole must have been aware of the fact that there are eight unknowns and, originally, as in Section 22, only six equations. Therefore, he must have known perfectly well that Cayley's solution had to have added two equations. Since one of the added equations was the algebraic expression of the assumed independence of the causes (A and B) it followed that there had to be another equation. As we will see in Section 28, Cayley was himself well aware of this.

It would have been very much to Boole's advantage to identify the other equation more adequately. It is certainly not possible to go from his verbal statement to a unique and equivalent algebraic equation. in his comments on Boole's numbered statements, Cayley apparently agrees with Boole: "I do assume that the causes A and B are absolutely independent of, and uninfluenced by each other; viz. not only the probability of A acting, but also the probability of its acting efficiently, are each of them the same whether B does not act, or acts inefficiently, or acts efficiently; and the like for B." Even with this very precise statement by Cayley it does not seem possible to deduce a corresponding unique equation.

The reason why it would probably have been to Boole's advantage to isolate the two equations representing what he believed Cayley was assuming in his solution is because of Boole's seventh statement. This is the heart of the issue and it is too bad Boole did not make the point more strongly. Precisely, Boole's system does encompass any possible set of assumptions, including Cayley's, in the sense that his system will produce the correct answer from any acceptable set of assumptions. in particular, if we apply Boole's method to the eight equations given as the basis of Cayley's solution (at the end of Section 23) we will get exactly Cayley's solution. Similarly, Boole could take either one of the two equations and add it separately to the original six and obtain the correct solution. Therefore, however important Cayley thought his distinction between the causal version and the concomitance version, both of the versions are encompassed by Boole's system.

This is worthy of illustration. Suppose we make only one of Cayley's assumptions, namely that A and B are independent. The six linear equations representing the previously given information are:

$$b = r - a$$

$$c = s - a$$

$$f = p - r - e$$

$$g = q - s - e$$

$$d = 0$$

$$h = 1 - p - q + a + e \ .$$

The single assumption of Cayley can be written

$$a + e = pq \ .$$

Solving for "e" and substituting in the above equations gives seven equations in seven unknowns:

$$e = pq - a$$

$$b = r - a$$

$$c = s - a$$

$$f = p - r - pq + a$$

$$g = q - s - pq + a$$

$$d = 0$$

$$h = 1 - p - q + pq \ .$$

We now need three new variables: t = AC, u = BC, and w = AB. By ordinary logical multiplication we get:

$$tuwAB = ABC = a \qquad t'u'w'A'B = A'BC' = g$$

$$tu'w'AB' = AB'C = b \qquad t'u'w'AB' = AB'C' = f$$

$$t'uw'A'B = A'BC = c \qquad t'u'w'A'B' = A'B' = d + h \ .$$

$$t'u'wAB = ABC' = e$$

Dividing the left-hand terms by the right-hand terms, setting each of them equal to each of the others, and setting each of them equal to V, the sum of the seven left-hand terms, gives seven equations in the seven unknowns t, u, w, A, B, V, and e. Multiplying the 2nd, 3rd, and 4th expressions and equating the product to the product of the 1st, 5th and 6th expressions gives equal numerators. Equating the denominators gives

$$bce = afg$$

or

$$(r - a)(s - a)(pq - a) = a(q - s - pq + a)(p - r - pq + a) \ .$$

This is a cubic in "a" and from "a" the whole logical table can be reconstructed. As usual, there is no need to determine the various Boolean variables.

Taking the same numerical illustrations used previously (p = .5, q = .6, r = .4, s = .42), the equation for "a" is

$$(.4 - a)(.42 - a)(.3 - a) = a(a - .12)(a - .2) \quad .$$

The solution of this equation is a = .246229, from which the logical table is reconstructed:

	A		A'	
	B	B'	B	B'
C	.246229	.153771	.173771	0
C'	.053771	.046229	.126229	.2 .

Comparison with Cayley's solution will show that the sums of the four columns (p(AB), p(AB'), and so forth) are identical, as they must be with the given assumption. However, the solutions are different because the other assumption of Cayley is not made. It needs to be emphasized again that if both assumptions are made then Boole's method will produce the same conclusion as Cayley obtained.

Of course, an alternative reading of Boole's first statement is possible. Perhaps he meant that Cayley's assumption of the independence of A and B might be reasonable but that no additional other assumption is equally reasonable. in this case it would have been even more effective had Boole worked out the solution just given, which incorporates this point of view. It is not at all easy to understand why Boole did not take this quite obvious step in presenting his side of the case.

My conclusion, then, is that Boole did not present his position nearly as effectively as he might have. With some obvious additions, such as the preceding example, Boole could clearly have carried the day against Cayley. This conclusion is reached, and all of the remarks of this Section are made, without reference to the criticisms of Wilbraham and a Note concerning these criticisms made subsequently by Cayley. This important contribution to the dispute will be discussed in Section 28.

Section 26. Keynes on the Challenge Problem.

Keynes, of course, refers to the Challenge Problem.[5] Keynes says: "After calculations of considerable length and great difficulty" Boole finds the desired p(C) as the root of a quadratic equation. Keynes then says: "This solution can easily be seen to be wrong." and adds "The mistake which Boole has made is the one general to his system ...". in Section 4 I have already commented about this quite ridiculous claim that Boole's definition of independence somehow invalidated his argument. But because of the prestige of Keynes it is perhaps desirable to see why Keynes believes that it is easy to see that Boole's solution is wrong.

Keynes argues thus. Suppose A and B cannot both occur. Clearly, then, the solution of the Challenge Problem is

$$p(C) = r + s \quad .$$

However, as Keynes correctly points out, Boole's solution does not reduce to this answer under the stated assumption.

All that is demonstrated by this argument is almost complete lack of understanding by Keynes of the elements of Boole's method. There is no reason why Boole's solution should reduce to the stated answer for very much the same reason that Boole's argument against Cayley's solution (discussed in Section 25) was not valid. in both cases the counter-example fails because it violates the conditions under which the solution was obtained. By taking the case where both A and B cannot occur, Keynes has conditioned, to use Boole's term, these variables. And Boole has very carefully stated that any such conditioning must be included among the equations describing the information known about the problem.

In the present instance this means that we must include the two equations a = e = 0 in the logical equations. This means that the equations expressing the known information become:

$$b = r$$

$$c = s$$

$$f = p - r$$

$$g = q - s$$

$$h = 1 - p - q$$

and

$$a = d = e = 0 \quad ,$$

It is clear that there are no unknowns and, of course,

$$p(C) = a + b + c + d = r + s \quad .$$

Whenever these equations are completely determined, as here, then Boole's method will produce the determinate answer as solution. But perhaps it would be well to demonstrate that fact here.

Introducing the two Boolean variables t = AC and u = BC, we have, by ordinary logical multiplication:

$$tuAB = ABC = a = 0$$

$$tu'AB' = AB'C = b$$

$$t'uA'B = A'BC = c$$

$$t'u'AB = ABC' = e = 0$$

$$t'u'AB' = AB'C' = f$$

$$t'u'A'B = A'BC' = g$$

$$t'u'A'B' = A'B' = d + h = h \ .$$

The resulting equations are:

$$\frac{tu'AB'}{b} = \frac{t'uA'B}{c} = \frac{t'u'AB'}{f} = \frac{t'u'A'B}{s} = \frac{t'u'A'B'}{h} = V$$

where V is the sum of the five numerators. Here there are five equations and five unknowns. I have not given the values of the Boolean variable generally because they are of no independent interest. Let us solve for them here. Inspection of the equations is sufficient to see quickly how to solve for each variable separately in the form A'/A or (1 - A)/A. Thus, from the 3rd and 5th of the equations

$$t'u'AB'/f = t'u'A'B'/h$$

so

$$A'/A = h/f$$

or

$$A = f/(f + h) \; .$$

Similarly,

$$B = g/(g + h)$$

$$t = b/(b + f)$$

$$u = c/(c + g) \; .$$

We can now multiply out each of the five terms entering into V and sum them to find

$$V = \frac{fgh}{(f + h)(g + h)(b + f)(c + g)}$$

the desired p(C) is

$$\begin{aligned} p(C) &= (tu'AB' + t'uA'B)/V \\ &= fgh(b + c)/fgh \\ &= b + c \; . \end{aligned}$$

This merely illustrates the fact that Boole's procedure will produce the unique solution when the logical equations have no unknowns, as here. It also serves to show that it is possible to determine the values of the Boolean variables if they are desired.

In any case, it is clear that the "counter-example" of Keynes is not at all a counter-example. Boole's method gives the correct solution when the problem is formulated correctly. Since the analysis is so simple, one is entitled, I believe, to doubt that Keynes had much understanding of Boole.

Section 27. Boole's Variations on the Challenge Problem.

In his Chapter XX ("Problems Relating to the Connexion of Causes and Effects"), in which he treats the Challenge Problem, Boole also analyzes three variations of this problem. These are worth considering as further examples of Boole's method and because they will be useful in the context of the discussion of Chapter IV of the present book.

The first such variation is his Problem 2 of Chapter XX. This is identical with the Challenge Problem except that Boole replaces the assumption that C cannot occur in the absence of both A and B. Instead of this assumption he assumes that A and B cannot both be absent. in other words, at least one of A and B must occur. The other conditions of the Challenge Problem remain unchanged in Problem 2.

We have, then,

$$d = h = 0$$

$$a + b + e + f = p$$

$$a + c + e + g = q$$

$$a + b = r$$

$$a + c = s$$

$$a + b + c + d + e + f + g + h = 1 \quad .$$

These equations can be solved in terms of one unknown:

$$b = 1 - q - p + r + e$$

$$c = 1 - q - p + s + e$$

$$f = p - r - e$$

$$g = q - s - e$$

$$a = p + q - 1 - e \quad .$$

In troducing the Boolean variables t = AC and u = BC and using ordinary logical multiplication gives:

$$\begin{aligned} tuAB &= ABC = a \\ tu'A'B &= AB'C = b \\ t'uA'B &= A'BC = c \\ t'u'AB &= ABC' = e \\ t'u'AB' &= AB'C' = f \\ t'u'A'B &= A'BC' = g \ . \end{aligned}$$

Following the procedure shown in Appendix C, we immediately find that

$$bce = afg$$

or

$$(1 - q - p + r + e)(1 - q - p + s + e)e =$$
$$= (p + q - 1 - e)(p - r - e)(q - s - e) \ .$$

This is a cubic equation in "e" and given "e", all of the other entries in the logical table are determined.

Boole, following his standard approach, deduces a cubic equation directly in u = p(C), the desired probability. This is always very much more difficult. The solutions, of course, are identical.

As a numerical example we can use the same known probabilities as in our previous illustrations (p = .5, q = .6, r = .4, and s = .42). Boole gives the "conditions for possible experience" for this example and these values happen to meet the requirements. The equation for "e" is:

$$(.3 + e)(.32 + e)e = (.1 - e)(.18 - e)(.1 - e) \ .$$

The root is e = .0123898 and the reconstructed logical table is:

	A		A'	
	B	B'	B	B'
C	.087610	.312390	.332390	0
C'	.012390	.087610	.167610	0 .

103

Here p(C) = a + b + c = .732390. This value, of course, satisfies Boole's cubic equation for u = p(C).

Problem 3 is exactly the same as the Challenge Problem but with a different question being asked. With the information being the same as for the Challenge Problem, Boole asks for the probability that event B occurs, given that event A occurs. in other words, he is asking for the ordinary conditional probability, p(B|A). This is an interesting example of the difference between the specific way Boole proceeded to apply his method and the procedure given here. Boole never referred to the underlying logical table and he also writes his equations using the desired probability as his unknown instead of determining how to reconstruct the logical table. in the present case this means that his solution of the

Challenge Problem is of no help in solving this problem and it means that for this problem he must go through his whole procedure again. It turns out that this is one of Boole's most onerous algebraic exercises and he needs several pages of argument to deduce the desired answer.

Obviously, from the logical table already reconstructed for the Challenge Problem it is easy to answer immediately any question based on the same information. This includes answering the present problem. Thus, directly from the reconstructed logical table for Boole's solution of the Challenge Problem (Section 24) we have p(AB) = .264640 and p(A) = .5 so p(A\B) = .529280. Naturally, this is exactly the same as the answer obtained by Boole.

It should be noted that Boole did not specify for this problem that p(A'B') = 0 and apparently did not intend this to be assumed. However, inspection of the Boolean argument leading to the solution of the Challenge Problem suffices to show that "d" and "h" cannot be separately estimated with the information given. in fact, this is the reason why Boole specified that p(A'B') = d = 0. Otherwise p(C) could not be exactly estimated. in Problem 3 the required answer does not depend on either "d" or "h" so it makes no difference whether we assume or not that p(A'B') = 0, the answer is the same.

The last example to be considered here is Boole's Problem 4. in this problem he assumes that in addition to the information of Problem 2 it is known that p(C) = m. He asks for p(AB).

The logical equations for this problem are the same in number as for the Challenge Problem because, so to speak, the condition that d = 0 in that problem is replaced by p(C) = m in Problem 4. This means that there are two unknowns. The equations can be written:

$$
\begin{aligned}
b &= r - a \\
c &= s - a \\
d &= p - r - e \\
g &= q - s - e \\
d &= m - r - s + a \\
h &= 1 - m - p - q + r + s + e \ .
\end{aligned}
$$

There are five Boolean variables and it is easy to deduce that

$$ad = bc \quad \text{whence} \quad a = rs/m$$

and

$$eh = fg \quad \text{whence} \quad e = (p - r)(q - s)/(1 - m) \ .$$

For a numerical illustration let us take p(C) = m = .6 in addition to the previously assumed values of p(A) = .5, p(B) = .6, p(AC) = .4, and p(BC) = .42. The reconstructed logical table is:

	A		A'	
	B	B'	B	B'
C	.28	.12	.14	.06
C'	.045	.055	.135	.165 .

This completes our analysis of Boole's various examples related to the Challenge Problem.

Section 28. Wilbraham's Criticism of Boole's Solution.

In our summary of the arguments concerning the solution of the Challenge Problem we have not yet discussed the most important criticism of Boole's solution. This was an article published by Henry Wilbraham in 1854.[6] This article annoyed Boole and, as Keynes puts it, he replied "hotly". One can hardly blame Boole for being annoyed. Wilbraham's style is not a particularly courteous one. Thus, in his summary of Boole's procedure he says: "...; that is to say, he solves a particular determinate case of an indeterminate problem, while his book may mislead the reader by making him suppose that it is the general problem which is being treated of." There are other remarks with similar implications so it is not surprising that Boole responded, after giving the above quotation: "I fear that the impression produced upon the mind of any person not acquainted with my work by such statements as the above would be, that I have introduced in a covert manner assumptions of the existence of which I was ignorant, or of the recognition of which I was afraid."[7]

Apart from this aspect of his style, one might wish that Wilbraham had not used some sixteen letters of the Greek alphabet in his discussion of some ordinary linear equations. He succeeds in making a basically simple argument quite difficult to follow. Further, he does not finally make his point in the direct way in which it could have been made. All of these various reasons may have contributed to the fact that Boole simply did not respond adequately to Wilbraham.

Obviously, however, such matters of style do not affect the correctness or not of Wilbraham's argument. What was his argument? in fact, his argument seems impeccable. Concerning the Challenge Problem he notes that there are eight unknowns and that the information given provides six equations. Therefore, he states clearly, the problem is indeterminate until two more equations are added by one or another assumption. He then undertakes to deduce the assumptions added by Cayley, on the one hand, and by Boole, on the other hand. He correctly deduces that both of them had one assumption in common and then he compares the other two assumptions. He considers Cayley's second assumption to be reasonable but of Boole's second assumption he says: "But the first of these equations appears to me not only arbitrary but eminently anomalous."

Boole's response to this criticism had, both here and elsewhere when his method was attacked, two aspects. First, he always insisted that his method never gave results not in accord with the "conditions of possible experience" of the data given. in his reply to Wilbraham, Boole made the point that Cayley's solution was more restrictive than the data required. (That this is so was shown in Section 25.) His second argument was, I am quite sure, based on his own profound conviction of the basic "rightness" of his system. in other words, his probability theory is erected four-square on his logical algebra and the two parts were, to Boole's mind, really the "Laws of Thought". Therefore, when someone claimed, as Wilbraham did, that his imputed assumptions were not reasonable, Boole had difficulty in replying because he genuinely believed that his method could only produce equations that were ineluctable deductions from the data by means of the very structure of thought. Thus, the closest he comes to replying to Wilbraham on this issue is: "They do not, indeed, represent `hypotheses', but they are legitimate deductions from the general principles upon which that method is founded, and it is to these principles directly that attention ought to be directed."[8] When Cayley challenged him on the same issue Boole replied: "No such explanation as you desiderate of the interpretation of the auxiliary quantities in my method of solution is possible; because they are not of the nature of additional data, and their introduction does not limit the problem as any hypotheses which are of that nature do. I do not see any difficulty whatever in the conception of this ideal problem."[9]

Despite the difficulty that Boole had in answering an attack, such as Wilbraham's, on the additional assumed equations that his method introduced, he surely gave thought to the question raised. in other words, I hypothesize that Boole must himself have found the additional equation isolated by Wilbraham somewhat difficult to understand. I am certain that it was this issue that was in his mind when he wrote in December, 1854: "These data exhibit both x and y as connected with z, and thereby also connected with each other. But that connexion is not of a kind which can be exhibited in an explicit form by means of propositions. And our consequent inability to express by any distinct and intelligible formula the implied relations among the elements x, y, z, renders it difficult to judge of the `reasonableness', or of the 'anomalous' character of results in the expression of which these elements are employed."[10]

Did Boole, then, have a "closed mind" on this basic question? I am sure that he was far too ardent a lover of truth for this to have been the case. However, it would have required an overwhelming argument to lead him to reconsider the beautiful structure of the "laws of thought" that he had uncovered. Such an argument was proposed and Boole definitely saw it. This was the final comment, almost an afterthought, of Arthur Cayley.

It will be recalled that there was quite a long interval of time between the original discussion between Cayley and Boole (1853-4) and the major exchange that took place in the article discussed in Section 25 (1862). Perhaps because of the lapse of time, Cayley forgot about Wilbraham's contribution and did not mention him in the 1862 article. Remembering, in the subsequent issue of the journal he added a Note in the form of a Postscript in which he summarized Wilbraham's argument.[11] His summary made Wilbrahman's point in a very much stronger way than Wilbraham had made it.

In this Note, Cayley briefly states the six known equations of the Challenge Problem. Then he states the two equations assumed for the two solutions. However, he states them in terms of the probabilistic variables involved. Thus, Cayley states, following Wilbraham, that both Cayley and Boole assume the truth of

$$ABC'/A'BC' = AB'C'/A'B'C' \ .$$

To make this more perspicuous, recall that for any two random variables that are independent it is the case that if $p(X) = p$ and $p(Y) = q$, then

$$p(XY)p(X'Y') = pq(1 - p)(1 - q) = p(X'Y)p(XY') = (1 - q)p(1 - p)q \ .$$

The relationship given by Cayley simply states that it is assumed that A and B fulfill this condition within C'. in other words, this assumption, made by both Cayley and Boole, is that A and B act as independent random variables within the population of C'. Cayley follows Wilbraham in stating that his own other assumption is that

$$(ABC + ABC')/(A'BC + A'BC') = (AB'C + AB'C')/(A'B'C + A'B'C') \ .$$

This equation can be written as

$$AB/A'B = AB'/A'B' \ .$$

Therefore, this equation means that Cayley is assuming that A and B behave as independent random variables over the entire universe (C + C'). Again, this is a plausible assumption.

Still following Wilbraham, Cayley now states the second assumed equation of Boole as

$$ABC/A'BC = AB'C/A'B'C' \ .$$

Now, each of the two preceding equations has an immediate interpretation in terms of the assumption of independence under one or another circumstance. This equation does not. Therefore, Cayley asks, What is it?

Making the point as directly as possible, Cayley says that since the equation does not reduce to an identity under the assumption that the variables involved are independent then Boole must explain where the equation comes from. Both of the two preceding equations did reduce to identities under the assumption that the variables involved are independent.

This appears to be an overwhelming argument, one that would apparently have forced Boole to re-examine his system. How can we ever know whether a system such as Boole's is true or not? in some sense, it is probably impossible for us ever to be sure that such a system is true. However, we certainly have a direct and reasonably simple way to determine whether it is false. This is to see whether it gives the correct answer in those cases where we can otherwise determine the correct answer. in dealing with probabilities we certainly can calculate the correct answer when we know that the events involved are independent. Therefore, a system such as Boole's absolutely must produce the correct answer when the events are independent.

This is why Wilbraham's argument, as sharpened by Cayley, appears to be so devastating for Boole's system. Since his system, for the Challenge Problem, introduces an implicit equation that is not true for independent variables it seems to follow immediately that the answers produced by his system will not be, at least in those cases, correct for independent variables. Therefore, the system will be, for at least those cases, unequivocally false. I repeat, this certainly **appears** to be the case.

Boole certainly saw Cayley's Note and he would have instantly understood the strength of the argument. He never replied to the argument. The Note appeared in the latter part of 1862. Boole died in 1864 at the tragically young age of 49. in the last two years of his life he had many duties and family responsibilities and was continuing his writing in several areas. Perhaps he simply did not have time to respond to the argument. But the issue raised was so absolutely central to Boole's major lifelong intellectual interests that one must imagine that he would have found time had he had an immediate answer available. There is, however, an alternative explanation: this is that Boole had tried several times to refute such arguments and had not succeeded in convincing his critics so he probably wanted to spend sufficient time preparing an answer so that he would have an adequate refutation of the argument. He was not given enough time.

I want here to emphasize what we will see in Chapter IV. THIS IS THAT DESPITE INITIAL APPEARANCES THE CAYLEY-WILBRAHAM CRITICISM IS WRONG. IT IS AN INVALID ARGUMENT. The remarkable structure that Boole created cannot be so easily overthrown, even by an argument that initially appears as strong as Cayley's and Wilbraham's.

Chapter IV of this book will be devoted to my attempt to answer this argument. But I wish to emphasize again that the various examples and uses of Boole's system given in Chapter II, and to be given in Chapter V, remain correct quite independently of this argument. Even if the argument were completely correct, it only applies in some instances that happen to include the Challenge Problem. There remain large areas where Boole's method is indubitable and works fully as well as Boole thought it did everywhere.

CHAPTER IV

BOOLE'S THEORY OF PROBABILITY:
SOME COMMENTS, A CORRECTION, AND AN EXTENSION

Section 29. Guide to This Chapter

We will begin this Chapter by discussing, in Section 30, the criticism of Boole's theory of probability initiated primarily by Wilbraham[1] but very much strengthened by Cayley[2]. A seemingly very strong argument, formulated mostly by Cayley, purports to show that Boole's solution to the Challenge Problem is incorrect. Of the two equations added to the known data of the problem by Boole's method, only one reduces to an identity if the data are taken to be of independent random variables. Cayley, in particular, argues that the one that does not so reduce must be incorrect and, therefore, the solution must be incorrect. There appears to be no escape from this conclusion and, hence, the argument has been generally considered to refute Boole's whole theory of probability. This is particularly unfortunate because the argument, in fact, is essentially incoherent, it does not in any way affect Boole's theory, and there is abundant evidence that, in particular, Boole was fully aware of the fact that the argument was irrelevant to his theory. We will undertake to demonstrate these statements in Section 30.

Boolean variables are ones for which we know, or can calculate, the probabilities. Therefore, the number of Boolean variables in a given problem can vary considerably, depending on the information we have available. Thus, suppose there are three simple events. Then if we know the probability of each we will have three Boolean variables. If we know the values of only two of the simple events and have to introduce two complex events then we will have four Boolean variables. There is generally no simple relationship between the probabilities of the simple and complex events in a problem and the values of the calculated Boolean variables. Thus, in Section 19 we considered an example with three simple events with probabilities: p(A) = .3, p(B) = .4, p(C) = .5. In addition, we introduced two complex events, s = AB with u = p(s) = .18 and t = BC with v = p(t) = .28. But when we calculate the values of the five Boolean variables we find them to be: p(A) = .2, p(B) = .178378, p(C) = .366667, p(t) = .801205, and p(s) = .765957. Nonetheless, it is a fact that with these values of the Boolean variables we can calculate exactly the probability of any cell in our logical table (see Section 31).

The relationship of the Boolean variables to the small-letter variables helps us in finding the relationships between denominators in the Boolean equations which enable us to solve for small-letter variables. Thus, a commonly discovered relationship is bh = df. Translating this into the underlying Boolean variables we get (AB'C)(A'B'C') = (A'B'C)(AB'C'). Removing the B' which occurs in each of the four terms we get (AC)(A'C') = (A'C)(AC') and this is a well-known relationship ordinarily verbalized as "A and C are random and independent variables within B'". There are six equations that express this kind of relationship. Another way to view bh = df is to write it as b = df/h and translating this into the underlying Boolean variables we get b = (A'B'C)(AB'C')/(A'B'C') = (AB'C) = b from which it follows that df/h is a valid way of expressing "b". This kind of argument leads to the conclusion that Boolean analysis is the same as that of the orthodox theory when the probabilities of the simple events are all known. In Section 31 we show that if we know the probabilities of the original simple events and follow Boole's procedure for any compound events defined over the simple events then we will always find random and independent variables. Thus, Boole's analysis will always produce estimates agreeing with the orthodox theory as applied to random and independent events.

In Section 31 we show the advantage of adding to Boole's "proceed from the simple to the complex" the powerful "proceed from the complex to the simple". This means that we should use the principle that probabilities of simple events should be assumed to be those probabilities necessary in order to produce the known joint probabilities under the assumption that the simple events are independent. This is the "extension" referred to in the title of this Chapter. A number of illustrations are given of adding an estimated probability of a simple event to the original data in order to avoid the difficulties that can arise when one or more such probabilities are unknown.

Section 32 is devoted to a question to which Boole ascribed considerable importance. This question was "Is a given Boolean analysis unique or are there alternative analyses that are also Boolean?" Boole's argument on the subject was based on his analyses of "the conditions of possible experience". He apparently believed that a valid analysis could be applied and would produce the correct result for any problem that met the requirements of "the conditions of possible experience". It would follow from this conceptualization that if two possible analyses were available then the one that had the larger conditions of possible experience would be the valid one. Boole used this argument in his disagreement with Cayley about the Challenge Problem but Cayley simply rejected the argument and Boole did not again advert to it in that context. However, Boole gave quite a lot of attention to this question with regard to his ingenious example of a group of people who speak a dialect in which there are no words "rain", "snow", "wind" but there are words "sleet" = both rain and snow, "drift" = both wind and snow, and "storm" = both wind and rain. Boole argued that the analysis of a problem should produce the same answer for either formulation. Section 32 shows, using an example, that this hope is not true and that an analysis in terms of rain, snow, and wind is very much preferable and that there simply is no equivalent analysis in terms of sleet, drift, and storm.

Section 33 is devoted to Boole's treatment of conditioned variables. There can be no question that Boole overlooked a problem here. To avoid excessive length, Section 33 is divided into three related parts: 33-A on Boole's Theory and Donkin's Principle; 33-B on Boole's Note, his most surprising mistake, and 33-C on Donkin's Principle and Conditioned Variables. Section 33-A makes clear that Boole ascribed very great importance to Donkin's principle and defines exactly when this principle can be used. Section 33-B discusses the surprising error made by Boole in which he corrected an example - apparently while his book was being published - and thereby changed a perfectly correct answer into a major error. His motivation, as far as it can be reconstructed, was to make his answer in accord with Donkin's principle. Section 33-C then shows that it is simply impossible to add an equation to account for a conditioned variable, as Boole seemed to believe. There are several ways in which a conditioned variable can affect a logical table and there is no single way to handle all of them. On the other hand, a conditioned variable **can** be correctly introduced into the logical equations but it must be done in terms of the specific effect of the conditioned variable.

Section 30. The Correctness of Boole's Theory

Because of the great importance ascribed to the Cayley-Wilbraham argument, let us briefly summarize the major steps of the argument. Boole claims that his method permits the solution of very many probability problems for which the orthodox theory does not give specific values although it can give bounds. His claim, of course, may be true or false. While we can never hope to prove that it is always true, we can perhaps show that it must be false. The theory can be taken to be false if it does not correspond to the results given by the orthodox theory when that theory does give an exact answer to a probability problem. Now, the orthodox theory will give an answer to any probability problem if each of the variables has a known probability and if all the variables are independent. Therefore, Boole's theory, or any other that might be proposed, must given the correct answer when these circumstances are true. But this means that any equation added by Boole's theory must reduce to an identity when the variables are random and independent. As Cayley points out, this is not true of one of the equations added by Boole's method when applied to the Challenge Problem. Therefore, the conclusion seems to follow: the method, at least in this instance, must be wrong.

This seemingly powerful argument is, in fact, quite irrelevant to Boole's solution of the Challenge Problem. The whole argument is so badly stated that it is not even coherent. A correct statement would go approximately as follows: if, and only if, a specific probability problem has a known solution by the orthodox theory then Boole's theory, or any other, must produce the same solution or else stand condemned as being false, at least for that probability problem. Now, there is NO solution to the Challenge Problem using the orthodox theory. Therefore, there is no way that Boole's theory can be criticized as not producing the answer of the orthodox theory - there is no such answer. Before defending this statement, it is well worthy of emphasis that Boole was completely aware of, and familiar with, the correct argument.

Boole gives a summary of the probability principles listed by Laplace that have been applied to the solution of questions of probability (LT, pp. 248-9). With regard to the extent and relative sufficiency of these principles he writes, "It is always possible, by the due combination of these principles, to express the probability of a compound event, dependent in any manner upon independent simple events whose distinct probabilities are given." An even more striking statement is also given, (LT, p. 250): "It may, therefore, be fully conceded that all problems having for their data the probabilities of independent simple events fall within the scope of received methods." It is beyond the bounds of rational opinion to imagine that the person who wrote these summary statements would then proceed to publish examples of solutions to probability problems which were not in accord with orthodox theory, granted that the requirements of the orthodox theory were available.

Boole states that it may be conceded "that all problems having for their data the probabilities of independent simple events fall within the scope of received methods". Let us therefore recall that the usual table of probabilities for three random and independent events, written in the usual format, is

	A		A'	
	B	B'	B	B'
C	a	b	c	d
C'	e	f	g	h .

Now, it is a fact that can be easily proven that for any such table, if the three events are random and independent, then it is true that bceh = adfg. Remember that the entries in the table (a, b, . . ., h) are the probabilities, observed or theoretical, of the corresponding combination of the three underlying variables, A, B. and C. To see that the equation is true, suppose that p(A) = p, p(B) = q, and p(C) = r. Then, if the three variables are random and independent we can calculate the probability of each of the eight entries in the table. Thus, b = p(AB'C) = p(1 - q)r, c = p(A'BC) = (1 - p)qr, e = p(ABC') = pq(1 - r), and

h = (A'B'C') = (1 - p)(1 - q)(1 - r). Taking the product of these four probabilities, we find that bceh = p2q2r2(1 - p)2(1 - q)2(1 - r)2. Repeat this simple analysis for adfg and you will find exactly the same expression. Therefore, for three random and independent events it is certainly true that bceh = adfg. Now, for the Challenge Problem we are given that p(d) = 0. Therefore, adfg = 0 but we are given no such information about bceh and there is no possible way for it to equal 0. Even if it were possible, the equation in the form 0 = 0 would be meaningless. Therefore, the three variables for the Challenge Problem do not meet the most basic requirement in order for them to be random and independent. Therefore, they are not random and independent. And therefore orthodox probability theory cannot deal with the Challenge Problem. It is worth mentioning that there are analogical equalities for four underlying variables or five or any larger number although it quickly becomes tiresome to write out the equality for larger numbers of variables.

Here is a numerical example. Suppose you are given the table:

	A		A'	
	B	B'	B	B'
C	.013682	.031096	.030100	.068410
C'	.081804	.185918	.179969	.409021 .

Are these three variables random and independent? The quickest way to find out is to verify whether bceh = adfg. We find bceh = .000031318 and adfg = .000031318 so our answer is affirmative. In fact, the table was constructed with p(A) = 5/16, p(B) = 11/36, and p(C) = .143288 and assuming that the three variables are random and independent.

However, this relationship is far more useful than it might now appear so a further example is justified. Suppose we are given the following information: p(A) = p = .5, p(B) = q = .6, p(C) = u = .55, p(AC) = r = .4, and p(BC) = s = .42 and we want to make a standard Boolean analysis of these data. Using our regular format (a, b, . . . , h) we know five pieces of information plus one because the sum of the eight cells equals one. Thus, we can write six equations that represent the known information. We can conveniently take "a" and "e" to be the two unknowns:

$$b = r - a \qquad d = u - r - s + a \qquad g = q - s - e$$

$$c = s - a \qquad f = p - r - e \qquad h = 1 - u - p + r - q + s + e\ .$$

(Notice that we can check our algebra at this point by remembering that a + b + c + d + e + f + g + h = 1.) We have information about two compound events so we will set t = AC, hence t' = A' + C', and w = BC, hence w' = B' + C'. We will obtain the Boolean variables by logical multiplication and it is convenient to do it in two steps. First, tw = ABC, t'w = A'BC, tw' = AB'C, and t'w' = A'B' + A'C' + B'C' + C'. Then it is straightforward to muliply by the eight possibilities involving A, B, and C and their negations. We quickly find:

$$twABC = ABC = a \qquad t'w'ABC' = ABC' = e$$
$$tw'AB'C = AB'C = b \qquad t'w'AB'C' = AB'C' = f$$
$$t'wA'BC = A'BC = c \qquad t'w'A'BC' = A'BC' = g$$
$$t'w'A'B'C = A'B'C = d \qquad t'w'A'B'C' = A'B'C' = h\ .$$

The Boolean equations are, as usual, obtained by dividing the left-hand side of each of these eight equations by the right-hand side and equating the eight expressions to each other and to V:

$$\frac{twABC}{a} = \frac{tw'AB'C}{b} = \ldots = \frac{t'w'A'B'C'}{h} = V$$

We can solve for the Boolean variables (A, B, C, t, w) but this takes more calculation than simply finding two equations (since we have two unknowns) from the denominators of the expressions in the equation that have, by multiplication, the same numerators. Thus, inspection discloses that the product of the 6th and 7th of these expressions has the same numerator as the product of the 5th and 8th. Therefore, we can equate the denominators of these two products and we immediately get eh = fg. Noting the pattern, we try the same pattern on the first four expressions and we find that the product of the 1st and 4th has the same numerator as the product of the 2nd and the 3rd. So, we equate the denominators and obtain ad = bc. We should notice the meaning of these two relationships. Writing the products in terms of A, B, and C we see that eh = fg becomes (ABC′)(A′B′C′) = (AB′C′)(A′BC′) or, simply, within C′ we see that A and B are random and independent variables, meaning that (AB)(A′B′) = (AB′)(A′B) within C′. The other equation we discovered, written out similarly, shows that A and B are random and independent variables within C.

From the first of these two equations we have, using our equations for the small-letter variables, that e(1 - u -p + r - q + s + e) = (p - r - e)(q - s - e) and from the second equation we have that a(u - r - s + a) = (r - a)(s - a). Solving these two equations we find

$$e = \frac{pq - ps - rq + rs}{1 - u}; \qquad a = \frac{rs}{u}$$

For our data we find that e = .04 and a = .305455. We can now use the six equations with which we started to fill out the complete table:

	A		A'	
	B	B'	B	B'
C	.305455	.094545	.114545	.035455
C'	.04	.06	.14	.21

It is easy to confirm that each of the data given initially are represented by the appropriate cells in this table. Thus, p(t) = p(AC) = a + b = .4 as it should, and similarly for p(w) = p(BC) = a + c = .42, and p(A) = a + b + e + f = .5, and p(B) = a + c + e + g = .6, and p(C) = a + b + c + d = .55. Finally, we can calculate that bceh = .000090969 and adfg = .000090971, equal within rounding errors. Therefore, the Boolean analysis has resulted in three random and independent variables that reproduce all the known information.

A seemingly natural assumption would then be that we ought to be able to calculate values for p(A), p(B), and p(C) that would reproduce the entries in the table. But this is impossible. As we know, there are five Boolean variables, not three: A, B, C, t, and w. We can, however, deduce the values of the five variables and they will, indeed, reproduce the values in the table. Equate the expressions for "e" and "g" and we find t′w′ABC′/e = t′w′A′BC′/g from which follows A′/A = (1 - p)/p = g/e and p = e/(e + g) =

.04/.18 = 2/9, so 1 - p = 7/9. Similarly, from other expressions we can find the values of the four other Boolean variables:

p = 2/9

q = .4

u = .144446

t = .903227

w = .828953 .

The sum of the eight numerators in our equation = V. Here we find that V = .03147148. Now, if we calculate the numerator for any cell in our table and divide the result by V we will obtain the exact entry in that cell. Thus, t'w'A'B'C = d = .00111579 and .00111579/.03147148 = .035454, the entry for "d", and t'w'A'BC' = g = .00440588 and .00440588/.03147148 = .14, the entry for "g". The same argument works for each cell in the table. So, even though we actually used five Boolean variables the fact still remains that the events underlying the variables are random and independent. Therefore, our test of this hypothesis, whether bceh = adfg, confirms that the variables are random and independent even though the number of variables is greater than what is shown by the table itself.

We said above that the Challenge Problem does not meet the most basic requirement for random and independent variables and that it therefore follows that orthodox probability theory cannot deal with the Challenge Problem. This statement is true because variable C is conditioned (Boole's term) in the sense that d = A'B'C = 0, meaning that C cannot occur in the absence of both A and B. This single value makes it impossible for adfg = 0 to equal bceh. Hence the underlying variables are not random and independent. Despite this fact, Cayley seemed to be convinced that the two equations he added to the six given for the Challenge Problem did, in fact, reduce to identities if the variables were random and independent. He also was convinced that Boole's two equations did not reduce to identities and from this Cayley concluded that Boole's solution was incorrect.

Let us review the reasoning involved. Both Cayley and Boole used the same first additional equation. This was that

$$\frac{ABC'}{A'BC'} = \frac{AB'C'}{A'B'C'}$$

To understand the meaning of this equation one needs to recall that if p(X) = p and p(Y) = q, and if X and Y are random and independent, then

p(XY)p(X'Y') = pq(1 - p)(1 - q) = p(X'Y)p(XY') = (1 - p)qp(1 - q) .

The equation that both of them used as the first additional one states that A and B fulfill this condition within C', meaning that A and B are independent random variables within the population of C'. Writing this same equation using the letters in our tabular presentation gives e/g = f/h or eh = fg.

In our immediately preceding example we discovered the same relationship by equating the denominators of the product of the 6th and 7th expressions to the product of the 5th and 8th, giving the equation eh = fg. For that example we therefore suggested that we try the same pattern for expressions 1st through 4th. We did so and discovered that the product of the denominators of the 2nd and 3rd equaled the product of the denominators of the 1st and 4th, giving ad = bc. The difficulty for the Challenge Problem is that this second equation cannot possibly hold because of the conditioning that makes d = 0. Now, so to speak, Cayley, and also Boole, would have been pleased to have the second equation as

$$\frac{ABC}{A'BC} = \frac{AB'C}{A'B'C} \quad or \quad \frac{a}{c} = \frac{b}{d}$$

meaning that the second added equation would say the same thing about A and B over the population of C. But neither person can use this equation as it is written because d = 0 makes the right-hand side of each of the equalities meaningless. Instead of the impossible relationship above, Cayley used

$$\frac{AB}{A'B} = \frac{AB'}{A'B'} \quad or \quad \frac{a+e}{c+g} = \frac{b+f}{d+h}$$

as his second equation, in the form a + e = pq, c + g = (1 - p)q, b + f = p(1 - q), and d + h = (1 - p)(1 - q). Doing the cross-multiplication we see that Cayley's second equation is pq(1 - p)(1 - q) = p(1 - q)(1 - p)q. By summing in this way Cayley certainly avoids having d = 0 in a denominator but he also avoids dealing with C as a third variable. The immediately preceding equation states that A and B are random and independent variables. So, on the one hand, Cayley can argue that his second equation reduces to an identity for random and independent variables but, on the other hand, the identity is for only two variables, not the three that are actually involved. Therefore, one cannot cite Cayley's solution as if it were the orthodox theory dealing with the Challenge Problem.

Boole's solution follows exactly the same lines as all of the applications of his theory do. Setting t = AC and u = BC we have four Boolean variables: A, B, t, and u. C is not a Boolean variable because we do not know the probability of C. The two equations that can be deduced from the Boolean equation are bc = a(d + h) and fg = e(d + h). Notice that since d = 0 it follows immediately that eh = fg, the first added equation, and a = bc/h, the second added equation. These equations are automatic results of Boole's probability theory. Cayley states the second assumed equation of Boole as being

$$\frac{ABC}{A'BC} = \frac{AB'C}{A'B'C'} \quad or \quad \frac{a}{c} = \frac{b}{h}$$

In view of the fact that Cayley also used d + h as denominator instead of d = 0, it is somewhat surprising that he demands to know whence this formula came from. The equation states that A and B function as random and independent variables within the population of C, except that since A'B'C = d = 0 as a result of conditioning, h = A'B'C' is used as the denominator instead.

The conclusion of this somewhat lengthy discussion is that the Cayley-Wilbraham criticism of Boole's analysis of the Challenge Problem is not valid. If the underlying variables in a problem are random and independent - meaning that the orthodox probability theory can construct an answer to the problem - then Boole's analysis will produce the same answer. The arguments for and against the Boolean theory must be posed in terms of the very many kinds of probability problems which cannot be handled by the orthodox theory but for which the Boolean theory produces exact answers. There is not some mysterious error pointed out by Cayley and Wilbraham. If the orthodox probability theory can construct an answer to a given probability problem then Boolean analysis will produce the same answer.

In addition, the Boolean analysis produces exact answers to very many probability problems for which the orthodox theory can only, at best, produce bounds on a desired probability.

Section 31. The Usefulness of Boolean Variables

Boolean variables are those for which we know the probabilities. Suppose we have three variables in our usual table: A, B, and C. The number of Boolean variables in the problem we would be considering may be only three, if we know the probability of each of A, B, and C; or it might be only two of these three if we do not know, say, the probability of C; or it might be four if we do not know the probability of C and we have to introduce simple variables for two compound events, say t = AB and u = AC. In every case, the Boolean variables are ones for which the probabilities are known, or, if not known, they can be calculated.

It is very important to remember that the calculated values of the probabilities of Boolean variables may not have any apparent direct relationship with the original known values. Thus, in our example of Section 19 we were given that p(A) = .3, p(B) = .4, p(C) = .5, p(AB) = u = .18, and p(BC) = v = .28 where we set s = AB and p(s) = u = .18 and t = BC and p(t) = v = .28. We have five Boolean variables and when we complete the calculations we find that p(A) = .2, p(B) = .178378, p(C) = .366667, p(t) = .801205, and p(s) = .765957. These seem to be quite far from the original values. However, they will exactly reproduce each of the values (of a, b, . . ., g, h) in our completed table. We need V = sum of the eight logically possible combinations of the five Boolean variables = .06371191. Then, for example, ABC = stABC/V = (.765957)(.801205)(.2)(.178378)(.366667)/.06371191 = .126, as given in the table. Similarly, each other value given in our completed table is exactly reproduced by these values of the five Boolean variables.

We have seen abundantly often that the small-letter variables (a, b, . . ., g, h) are extremely useful and enable us to make the standard Boolean analyses with very much less difficulty than Boole found. The point of the present Section is to show that it is also quite helpful to remember the relationship of the small-letter variables to the underlying Boolean variables. In our example of Section 19 we set up the Boolean equations and each of the eight terms, set equal to each other and to V, had a single one of the small-letter variables as denominator. Our procedure was to look for two pairs of the individual terms which, when multiplied together produced identical numerators. When this occurs we know that we are entitled to equate the denominators. If we can do this twice then we will have two equations, which will permit us to solve for the two unknowns, "b" and "e". Knowing these, we can calculate every entry in the logical table.

We found, first, by observation that the term with denominator "b" times the term with denominator "h" had the same numerator as the term with denominator "d" times the term with denominator "f". Therefore, we know that bh = df and substituting the equations for these small letters gave us b(.26 + b) = (.22 - b)(.12 - b) from which we find b = .044. How did we find this relationship? By dint of long looking, or did we have some kind of help? The answer is that we had some kind of help - specifically that which results from converting the small-letter variables to their equivalent Boolean variable form. Thus, b = AB′C and h = A′B′C′, while d = A′B′C and f = AB′C′. Notice that there is a B′ in each term. Removing the B′ and rewriting we have (AC)(A′C′) = (A′C)(AC′). This is the well-known relationship that is verbalized as A and C are random and independent within B′. When we discover such a relationship within B′ it is indeed well worth investigating whether it is also true that A and C function similarly within B, as suggested in our discussion of this example in Section 19. If so, then we ought to find that (ABC)(A'BC') = (A'BC)(ABC'). This is true for ag = ce. (Notice that we go directly from the specified Boolean variable equation to the small-letter formulation since, for example, ABC = a.) This gives us the other required equation and we can now complete the logical table.

There are three such pairs that are logically possible:

1.	A and B random and independent within C:	(ABC)(A'B'C) = (AB'C)(A'BC) or ad = bc
2.	A and B random and independent within C':	(ABC')(A'B'C') = AB'C')(A'BC') or eh = fg
3.	A and C random and independent wit`hin B:	(ABC)(A'BC') = (A'BC)(ABC') or ag = ce
4.	A and C random and independent within B':	(AB'C)(A'B'C') = (A'B'C)(AB'C') or bh = df
5.	B and C random and independent within A:	(ABC)(AB'C') = (AB'C)(ABC') or af = be
6.	B and C random and independent within A':	(A'BC)(A'B'C') = (A'B'C)(A'BC') or ch = dg.

Thus, the example of Section 19 has 3 & 4; Boole's problem #4 from his Chapter XX, discussed in Section 21, has 1 & 2; Boole's solution of the Challenge Problem has 1 & 2, but with "h" instead of "d"; and Boole's problem 4, a variation of the Challenge Problem, has 1 & 2.

There is no requirement that a given problem must fulfill one or more of the above possible conditions. However, the indicated equations do arise quite often. They also do not have to occur in pairs. For example, Boole's Problem V, discussed in Section 21, fulfills 6, 4, and 2. These mean that A and B are random and independent within C', A and C are random and independent within B', and B and C are random and independent within A'. Finally, other equations can also be found in various probability problems. An example is the equation that holds if all three events are random and independent: bceh = adfg, where each side of the equation has $A^2(A')^2B^2(B')^2C^2(C')^2$. Knowing that one of these equations may hold makes it quite a bit easier to find an equation among the small-letter variables.

Let us now consider another aspect of the Boolean variables that is also very useful. Returning to the example used in Section 19, we may notice that from the equation bh = df we have immediately that b = df/h. Writing out the small-letter variables in terms of the Boolean variables we have

$$b = \frac{(A'B'C)(AB'C')}{A'B'C'} = AB'C = b$$

We see that which a moment's thought would suggest, namely that the estimator for one of the small-letter variables must be defined by Boolean variables that are in agreement with those of the definition of the given small-letter variable. Here we see that the estimator for "b" has exactly the same Boolean variables as the definition of "b". Similarly, from the second equation, ce = ag we have the estimator for "e" as e = ag/c and

$$e = \frac{(ABC)(A'BC')}{A'BC} = ABC' = e$$

Again, the estimator for "e" has the same Boolean variables as does the definition of "e".

These interrelationships between valid estimators and the definition of the estimated small-letter variable in terms of the Boolean variables provide an opportunity to avoid some difficulties that can arise in the Boolean analysis of questions about probabilities. Let us begin by noting that no difficulties arise in the Boolean analysis as long as all of the simple events have known probabilities. As usual with Boole's argument, examples that might be anticipated to be easy ones often turn out to be much more difficult than expected. Therefore, let us begin with the easiest possible illustration. Suppose we have just two events, A and B, and that we know p(A) = p and p(B) = q. Further, A and B are independent, random events and nothing further is known. In this simple case the logical table is

	A	A′
B	a	b
B′	c	d

Our known equations are: $a + c = p$ or $c = p - a$
$a + b = q$ or $b = q - a$
and $a + b + c + d = 1$ or $d = 1 - q - p + a.$

Since we do not have any information other than that stated concerning any entry in the logical table it follows that we do not need any compound event. The Boolean equations are:

$$\frac{AB}{a} = \frac{A'B}{b} = \frac{AB'}{c} = \frac{A'B'}{d} = V$$

V = 1 so each of the numerators of the four first fractions equals its denominator. Suppose p = .4 and q = .3. Then our equations become: c = .4 - a, b = .3 - a, and d = .3 + a. From A'B = .3 - a and A'B' = .3 + a we immediately obtain A' = .6; whence A = .4. Similarly, we find B = .3 and B' = .7. We have, then, that a = .12 and we see that every entry in our logical table is simply the same one we would calculate, knowing p and q and knowing that the events are random and independent. Thus, Boole's method recapitulates the orthodox theory in this case where the two events are random and independent. Reviewing the reasoning is sufficient to conclude that the same result will hold for any number of events, as long as they are random and independent. Notice that the numerators of the fractions are written, as always, in terms of the Boolean variables - here A, B, A', and B' - and that these are required, by the way they are written, to give the numerators multiplicatively.

It is worth pointing out that even in this simplest of all examples we have two quite different ways to approach the resulting desired analysis. The way we have used in the examples of Chapters 2 and 3 is to reason in terms of the denominators of the Boolean equations. This results in a very great simplification of Boole's argument in more complex examples. Here we would note that the product of the 1st and 4th Boolean equations has the same numerator as the product of the 2nd and 3rd: in fact, multiplying each of these two pairs together gives

$$\frac{ABA'B'}{ad} = \frac{ABA'B'}{bc}$$

whence ad = bc or a = bc/d = (q - a)(p - a)/(1 - q - p + a), which immediately solves to a = pq. Knowing "a", all four entries in the logical table are known and the analysis is completed. In following this route we essentially ignore the values of the Boolean variables: A, B, A', and B'. Once we have the values of a, b, c, and d we can quickly solve for the Boolean variables if we want to do so. Thus, we have AB = .12 and AB′ = .28. Adding the two equations gives A(B + B′) = A = .4 and we find similarly that B = .3. We could, instead, determine the values of the Boolean variables directly. Thus, from the equality of the first and second Boolean variables we can write

$$\frac{b}{a} = \frac{A'B}{AB}$$

or b/a = A′/A = (1 - A)/A, whence A = a/(b + a) and A′ = b/(b + a). Similarly, B = a/(c + a) and B′ = c/(c + a). Boole's V is the sum of the numerators of the four Boolean variables and it is easy to verify here that V = 1. (This, of course, is not always, or even generally, true.) Suppose we have the logical table with a = .12, b = .28, c = .18, and d = .42. Then we have immediately that p(A) = p = .12/.4 = .3, whence p(A′) = .6, and p(B) = .12/.4 = .3, whence p(B′) = .7. Notice that since A, B, and generally so forth, are simply the names of events that it follows that there is no ambiguity if we do as Boole often, or usually, does and let A, B, and so forth, taken arithmetically rather than logically, be an alternative notation for p(A), p(B), and so forth. Thus, the equations we have written in this paragraph for A, A′, B, and B′ are actually the equations for p(A), p(A′), p(B), and p(B′).

Let us illustrate the same fact - namely, that the Boolean analysis will produce the same results as does the orthodox theory when the probabilities of the simple events are known - with three events, random and independent. Suppose we have three independent random events, A, B, and C, and that Prob(A) = p, Prob(B) = q, and Prob(C) = r are known and that there is no further information. The desideratum is Prob(ABC). Then the logical table is

	A		A′	
	B	B′	B	B′
C	a	b	c	d
C′	e	f	g	h

and the known information is expressed in four equations:

$$a + b + e + f = p$$
$$a + c + e + g = q$$
$$a + b + c + d = r$$
$$a + b + c + d + e + f + g + h = 1\,.$$

We can write one form of the solution of these equations thus:

$$a = q + r + p + d + f + g - 2 + 2h$$
$$b = 1 - q - d - f - h$$
$$c = 1 - p - d - g - h$$
$$e = 1 - r - f - g - h\,.$$

Here we have simply solved for the events in the order that they occur alphabetically.

The Boolean equations are:

$$\frac{ABC}{a} = \frac{AB'C}{b} = \frac{A'BC}{c} = \frac{A'B'C}{d} = \frac{ABC'}{e} = \frac{AB'C'}{f} = \frac{A'BC'}{g} = \frac{A'B'C'}{h} = V$$

Our usual procedure - and one that saves a great deal of algebra as compared to Boole's approach - is to argue in terms of the denominators, the individual entries in the logical table. Thus, in this example we would argue that all we want is "a" (the stated desideratum) so we need two expressions with the same numerators and with "a" being the denominator of one of the two. It is easy to see that 4*5 = 1*8 is satisfactory. The two denominators are ah = de, whence a = de/h. Since we do not have equations for each of the eight logical quantities we cannot use this equation to immediately solve for "a". However, we can more naturally reason in terms of the Boolean variables in the numerators of the fractions. Here we know that the entire logical table could be filled out by calculating the numerators in each of the eight fractions above. Thus, d = A'B'C = (1-p)(1-q)r, e = pq(1-r), and h = (1-p)(1-q)(1-r). Therefore, a = pqr, as we knew it would by ordinary probability theory. The calculations resulting from the Boolean equations are exactly the same as we would make knowing only that we are dealing with independent random variables. Thus, if p = .3, q = .4, and r = .35 then, for example, A'B'C' = .7*.6*.65 = .273 = h, just as we would calculate it using orthodox theory. And the argument obviously holds true for any number of random events with the only requirement being that we know the probabilities of each of the simple events.

Suppose we were given the logical breakdown for some observed data and we want to estimate p(A) = p, p(B) = q, and p(C) = r in order to determine whether, in accordance with the model under consideration, these quantities are sufficient to determine the entire logical table. How would we try to estimate the three quantities? in doing this we would work primarily with the numerators of the Boolean equations. Thus, equating the 1st and 3rd of the Boolean equations we would have

$$\frac{ABC}{a} = \frac{A'BC}{c}$$

whence c/a = A'BC/ABC = A'/A and A = a/(a+c). Since we are assuming that we have the entire logical table we can immediately calculate A. Suppose we have the table

	A		A′	
	B	B′	B	B′
C	.042	.098	.063	.147
C′	.078	.182	.117	.273

Then A = .042/.105 = .4 (using, as Boole does, A as an alternative symbol for p(A)). Similarly, from the first and second of the Boolean equations we find B = a/(a+b) = .3 and from the first and fifth we obtain C = a/(a+e) = .35. Finally, suppose we wonder about the correctness of our deduced formulas. For example, can A really, meaningfully equal a/(a+c)? We can answer this question, and any others like it, most easily by referring to the Boolean variables for "a" and "c":

$$\frac{a}{a+c} = \frac{ABC}{ABC + A'BC} = \frac{ABC}{(A'+A)BC} = \frac{ABC}{BC} = A$$

We see that our equation does, indeed, give us "A".

What happens if we add some further information, but still requiring that we know the probabilities of each of the simple events? Let us return to our example of only two random events. Now we will suppose that the information about events A and B, still random but now not necessarily independent, remains the same. We assume that now we have additional information to the effect that AB = k, meaning that the outcomes of A and B will be correlated. (The outcomes are independent if k = pq.) Since we now have information about a compound event we must first introduce another single-letter event to correspond to AB. Suppose we set t = AB, whence t′ = A′ + B′. The effect of this change is to supply another variable so that we will be able to solve the Boolean equations. We would now have to write the logical table as

	A		A′	
	B	B′	B	B′
t	a	b	c	d
t′	e	f	g	h

Let Prob(t) = k and our information gives the following equations:

$$a + b + e + f = p$$
$$a + c + e + g = q$$
$$a + b + c + d = k$$
$$a + b + c + d + e + f + g + h = 1 .$$

By ordinary logical multiplication we find that ABt = AB = a; AB′t = b = 0; A′Bt = c = 0; A′B′t = d = 0; and ABt′ = e = 0, where the 0's are all logically impossible combinations; AB′t′ = AB′ = f; A′Bt′ = A′B = g; and A′B′t′ = A′B′ = h. Therefore, the Boolean equations are:

$$\frac{ABt}{a} = \frac{AB't'}{f} = \frac{A'Bt'}{g} = \frac{A'B't'}{h} = V$$

Returning to the equations that express our information, and taking account of the entries that we now know = 0, the equations can be written:

$$a = k \qquad g = q - k$$
$$f = p - k \qquad h = 1 - p - q + k$$

(Putting k = .12 gives, as it should, exactly the same Boolean equations that we had when we assumed that A and B were both random and independent.) We solve for A, B, t, and k by the usual procedure. Thus, using the 2nd and 4th Boolean equations we obtain A′/A = h/f. Using the values immediately above we obtain A = (p-k)/(1-q). Proceeding similarly for B, t, and V, we obtain, after substituting the values in the immediately preceding equations:

$$A = (p - k)/(1 - q)$$

$$B = (q - k)/(1 - p)$$

$$t = k(1 - q - p + k)/(k(1 - q - p + k) + (p - k)(q - k))$$

and

$$V = ABt/k\,.$$

So, we find that the effect of adding an additional piece of information is to change the values of the probabilities of the simple events but each entry in the logical table is still found by multiplying the probabilities together in exactly the same way as the orthodox theory does if the simple events are both random and independent. So to speak, because of the additional information we have to estimate the probabilities of the simple events but once this is done we calculate the probabilities of joint events exactly as with the orthodox theory.

As a numerical illustration, suppose $p = .4$ and $q = .3$, as in our two-variable example above. This time we will assume that $k = .15$, rather than .12 above. We immediately calculate that $A = .357143$, $B = .25$, $t = .642857$, and $V = .382653$. The four non-zero entries in the logical table are $a = .15$, $f = .25$, $g = .15$, and $h = .45$. Then, $ABt/k = V$, $AB't'/f = V$, $A'Bt'/g = V$, and $A'B't'/h = V$. For this kind of calculation Boole often writes the equations slightly differently: $ABt/V = k = .15$; $AB't'/V = f = .25$; $A'Bt'/V = g = .15$; and $A'B't'/V = h = .45$. We see that the entries in the logical table are proportional to the products of the probabilities of the three simple events. Simply calculating the numerators of the Boolean equations produces, in order, .057398, .095663, .057398, and .172194. The sum of these numerators is Boole's V. The additional step required is to normalize the calculated entries in the logical table by requiring that their sum should equal one, which means to divide each of the calculated numerators by V.

As a second example we will take the Challenge Problem, changed so that p(C) is also known. This change eliminates the need for the "conditioning" in the statement of the problem, as given in Chapter III. The conditioning was that C could not occur unless at least one of A and B occurred. This is not needed in the present version. We assume, then, that $p(A) = p$, $p(B) = q$, $p(C) = n$, $p(AC) = r$, and $p(BC) = s$. Since we have two compound events we must introduce single variables for each of them. We will let $t = AC$, hence $p(t) = r$, and $u = BC$, hence $p(u) = s$. This means that we now have five simple events. The logical table would have 32 cells and is too big to write out - particularly since a majority of the entries would be zero. Specifically, there will be eight non-zero entries, all of the possibilities in terms of A, B, and C and their negations. Let us discover the non-zero entries by using logical multiplication. It is easiest to do in two steps. First, we have $t' = A' + C'$ and $u' = B' + C'$. Therefore, we can quickly find:

$$\begin{aligned} tu &= ABC \\ tu' &= AB'C \\ t'u &= A'BC \\ t'u' &= A'B' + C'\,. \end{aligned}$$

We can now determine the non-zero possibilities by multiplying the eight possibilities in terms of A, B, and C by the above four. We quickly find:

$$\begin{array}{llll} ABCtu = ABC = a & & ABC't'u' = ABC' = e \\ AB'Ctu' = AB'C = b & & AB'C't'u' = AB'C' = f \\ A'BCt'u = A'BC = c & & A'BC't'u' = A'BC' = g \\ A'B'Ct'u' = A'B'C = d & & A'B'C't'u' = A'B'C' = h \end{array}$$

The equations representing our knowledge are:

$$\begin{aligned} a + b + e + f &= p \\ a + c + e + g &= q \\ a + b + c + d &= n \\ a + b &= r \\ a + c &= s \\ a + b + c + d + e + f + g + h &= 1\ . \end{aligned}$$

We have six equations and eight unknowns. We can write all of the unknowns in terms of the knowns and "a" and "f", giving

$$\begin{aligned} b &= r - a \\ c &= s - a \\ e &= p - r - f \\ g &= q - p - s + r + f \\ d &= n - r - s + a \\ h &= 1 - n - q + s - f\ . \end{aligned}$$

There are eight Boolean variables, each equal to V:

$$\frac{ABCtu}{a} = \frac{ABC't'u'}{e} = \frac{AB'Ctu'}{b} = \frac{AB'C't'u'}{f} = \frac{A'BCt'u}{c} = \frac{A'BC't'u'}{g} = \frac{A'B'Ct'u'}{d} = \frac{A'B'C't'u'}{h} = V$$

We could solve these equations for A, B, C, t, and u, which is ordinarily Boole's approach but which is very considerably more difficult than working primarily with the denominators. If we know "f" and "a" then we know the entire logical table so we would like two equations that would enable us to solve for "f" and "a". After some inspection we notice that the numerator of 4th * 6th equals the numerator of 2nd * 8th : $AA'BB'(C')^2(t')^2(u')^2$. Therefore, the denominators give fg = eh, whence f = eh/g. We notice that this result amounts to arranging the last four letters (all of the C' values) as fg = eh so we try the same arrangement with the first four letters (all of the C values), suggesting ad = bc, so equivalently, 1st * 7th and 3rd * 5th. We quickly see that the numerators of these products each = $AA'BB'C^2tt'uu'$. Therefore, from the denominators we obtain ad = bc, whence a = bc/d. We have the desired two equations. Substituting the values of the variables in terms of "a" and "f" we quickly find:

$$f = \frac{(p - r)(1 - n - q + s)}{1 - n} \quad \textit{and} \quad a = \frac{rs}{n}$$

Therefore, we are done since we can immediately construct the logical table for an example of this type.

As a numerical example, suppose that p = .4, q = .3, n = .35, r = .2, and s = .14. We calculate a = .08 and f = .150769. Using our equations for the other six variables in terms of "a" and "f", we get

	A		A′	
	B	B′	B	B′
C	.08	.12	.06	.09
C′	.049231	.150769	.110769	.339231

We note two things about this example. First, every entry in the logical table is unambiguously known so we can make estimates of either of the two probabilities that were missing from our original knowledge:

ABC and AB. Neither of these quantities can be estimated by orthodox theory. Second, as we see from the Boolean variables, every entry in the logical table is calculated multiplicatively from the five probabilities of the simple events: A, B, C, t, and u. Therefore, the Boolean analysis has certainly resulted in random and independent variables. As usual, we do not need to calculate these probabilities in order to reconstruct the logical table. However, we easily **can** calculate them if desired. Thus, from the 4th and 6th equations we can find A = e/(e+g) = (p-r-f)/(q-s); from the 2nd and 4th we find B = e/(f+e) = (p-r-f)/(p-r). Proceeding in this way, we obtain the following estimates of the probabilities of the simple events:

A = .307694
B = .246155
C = .209679
t = .75
u = .671232 .

We can now calculate that a = ABCtu = .007995, b = AB'Ctu' = .011992, and so forth until h = A'B'C't'u' = .033901. The sum of these eight quantities is V = .1 so the entries in the logical table are the preceding calculated quantities divided by .1.

These two examples suffice to illustrate the fact that if we know the probabilities of the original simple events and follow Boole's procedure for any further knowledge - hence, of compound events defined in terms of the original simple events - we will always find random and independent underlying variables. Boole's analysis will always produce estimates that will agree with the results if we start with independent, random variables and apply orthodox theory.

There are large classes of examples for which the Boolean analysis will find the simple events are random and independent variables. When, then, may it be otherwise? We have already stated the answer and we will repeat it here. The Boolean analysis will not find random and independent variables when at least one of the probabilities of the original simple events is not known. This result can also occur because of what Boole meant by "conditioning". In this Section we will be primarily interested in showing the effect of unknown probabilities of simple events on this kind of result.

Let us take as an example the Challenge Problem - but without the "conditioning" that Boole introduced in this problem - for reasons that will become clear. We will assume, then, that we know p(A) = p, p(B) = q, p(AC) = r, and p(BC) = s. Notice that we are assuming that we do **not** know p(C), one of the three simple events. We proceed as usual. We will set t = AC, so p(t) = r, and u = BC so p(u) = s. We have t' = A' + C' and u' = B' + C'. It is easy to find, by logical multiplication, that

tu = ABC
tu' = AB'C
t'u = A'BC
t'u' = A'B' + C' .

With these in hand we can easily find the following, also by logical multiplication, where we include C even though we do not know p(C) because our logical table will be over A, B, and C:

ABCtu = ABC = a
ABC't'u' = ABC' = e
AB'Ctu' = AB'C = b

$AB'C't'u' = AB'C' = f$
$A'BCt'u = A'BC = c$
$A'BC't'u' = A'BC' = g$
$A'B't'u' = A'B'(C + C') = d + h$.

Our information is summarized in these equations:

$$a + b + e + f = p$$
$$a + c + e + g = q$$
$$a + b = r$$
$$a + c = s$$
$$a + b + c + d + e + f + g + h = 1 .$$

We already know from the structure of our logical table (from the logical multiplication, above) that we will not be able to separately evaluate d and h. We have seven variables and five equations so we can solve the equations in terms of "a" and "f", for example:

$b = r - a$
$c = s - a$
$e = p - r - f$
$g = q - p - s + r + f$
$h + d = 1 - q - r - f + a$.

We will consider the estimation of p(C) and of p(ABC). It is already perfectly clear that we will not be able to obtain a specific estimate for p(C). This is true because d is part of p(C) and h is part of p(C') and we have no way to separately estimate d and h.

The Boolean equations are written as usual except that we do not include C as a Boolean variable since we do not know p(C):

$$\frac{ABtu}{a} = \frac{AB'tu'}{b} = \frac{A'Bt'u}{c} = \frac{ABt'u'}{e} = \frac{AB't'u'}{f} = \frac{A'Bt'u'}{g} = \frac{A'B't'u'}{d+h} = V$$

working with the denominators, we would like two products with the same numerator and with "a" in the denominator and another such with "f" in the denominator. We find that the numerator of the product of the 1st and 7th equals the numerator of the product of the 2nd and 3rd: AA'BB'tt'uu'. Hence, we have immediately that a(d+h) = bc or a = bc/(d+h). Similarly, the numerator of the product of the 4th and 7th equals the numerator of the product of the 5th and 6th: $AA'BB'(t')^2(u')^2$. This gives f = e(d+h)/g. Substituting our algebraic expressions for these logical classes we obtain:

$$a = \frac{(r-a)(s-a)}{1-q-r-f+a} \quad and \quad f = \frac{(p-r-f)(1-q-r-f+a)}{q-p-s+r+f}$$

The expression for "a" involves "f" and the expression for "f" involves "a". The best solutions I have been able to find give a quadratic for "a" and then a simple fraction in "a" for "f":

$$A^2(1 - q - p + r + s) + a[p(q + r - 1) + r(1 - q - r) - rs - (1 - q + s)(s + r - 1)] - rs(1 - r - s) = 0$$

and

$$f = [a(1 - q + s - p + r) + p(q + r - 1) + r(1 - q - r) - rs]/(s + r - 1)\ .$$

As a numerical example we will take the same values as we used in the preceding example at the end of Section 26: p = .4, q = .3, r = .2, and s = .14. Of course, in the present example we do not assume, as we did in the previous one, that we know p(C). We immediately calculate that a = .040837 and f = .154340. Then our equations for the other logical classes give:

b = .159163
c = .099163
e = .045660
g = .114340
d + h = .386497

and V = 1. In other words, the logical table is

	A		A′	
	B	B′	B	B′
C	.040837	.159163	.099163	.386497
C′	.045660	.154340	.114340	

We see that despite the missing information about p(C), we are able to make exact estimates, for example, of p(ABC) = a = .040837 and of p(AB) = a + e = .086497. However, the best estimate that we can make for p(C) is .299163 (the sum of the three entries in the row for C) + .386497k, where 0 ≤ k ≤ 1, meaning that p(C) can include an unknown proportion of d+h.

None of these estimates can be found with the orthodox theory. However, this statement - perfectly correct though it is - has nothing to do with whether or not Boole's method produces an estimate for a given example which does not correspond to the estimate of the orthodox theory under the assumption of random and independent variables. This is the case for the present example and it is entirely because p(C) is not known. In this example the effect of our ignorance of p(C) seems to be that it makes it impossible to separately estimate d or h. But the reality is much worse: every single one of the estimates for the logical classes in the table immediately above diverges from the estimates of the orthodox theory.

We can see that this statement is true by anticipating a result that will be demonstrated in Section 29 and by comparing our estimates for the present example with the estimates for the same example but knowing p(C), which we developed in Section 26. Suppose we actually have three events that are random and independent. Then, as usual, we would have the logical table

	A		A′	
	B	B′	B	B′
C	a	b	c	d
C′	e	f	g	h

and each entry would be calculated in the usual way by orthodox theory. Thus, a = ABC, d = A'B'C, and so forth. Now, it is a fact that can be easily verified that

$$\frac{bceh}{adfg} = \frac{AB'C*A'BC*ABC'*A'B'C'}{ABC*A'B'C*AB'C'*A'BC'} = 1$$

This relationship holds true for any three random and independent events by orthodox theory and, hence, it must hold true for any result of Boolean analysis of three events. If it does not hold true then the estimates from the Boolean analysis will certainly diverge from the orthodox estimates - if the orthodox estimates can be achieved with the given information. It is easy to check the completed logical table for the example at the end of Section 26 and it will be found that the above relationship is satisfied. However, in our most recent example - the same one as in Section 26 but without assuming that we know p(C) - we find that the relationship cannot be satisfied because we do not have d and h separately. in fact, however, the orthodox theory cannot be applied precisely because P(C) is not known. Therefore, here, as in all other similar cases, the Cayley-Wilbraham criterion cannot be applied because there is no way to apply orthodox theory.

We can take the given relationship and by solving it for any one of the eight variables we can immediately discover what must be the relationship that exists between that variable and the others so that the estimate of the given variable would not diverge from that of the orthodox theory. Thus, in our current example - not knowing p(C) - every other one of our estimated logical classes depends on "a". Solving the equation for "a", we see that the relationship should be a = bceh/dfg in order to avoid this divergence. Since we do not know d and h we are unable to meet this requirement. In fact, our estimate for "a" involves the root of a quadratic equation in "a" and the quadratic equation does not bear the slightest resemblance to the equation just quoted. Since our estimate of each logical class directly involves "a" we are surely entitled to summarize by saying that every estimate resulting from the Boolean analysis of this example would differ from the estimates of the orthodox theory if P(C) were known but without P(C) the orthodox theory cannot be applied.

The reason for the failure of the Boolean analysis to parallel the orthodox theory with random, independent events is that the probability of one of the simple events, p(C), is not known. Generally, if at least one simple event's probability is not known then there are at least two cells in the logical table that cannot be separately estimated. This means that in the logical equations the sum of these two cells enters as a single unknown. This, in turn, means that at least one equation added by Boole's method must involve this sum. Whatever equation does involve this sum will generally produce a divergent estimate as compared to the orthodox theory. This is illustrated in the present example by the fact that we cannot do better than to estimate the sum, d+h, rather than the the two individual logical classes. But the general relationship between, for example, "a" and the seven other logical classes is supposed to have "h" in the numerator and "d" in the denominator. This is impossible to achieve since we do not separately know "d" and "h". Hence the divergence in question arises.

What might be done about this difficulty with the Boolean analysis? Within Boole's system it is obvious that nothing can be "done" about the problem since it arises within the system. This granted, one can certainly ask whether it is possible to verbalize the reason the difficulty arises. I believe that the reason can be verbalized and that it is useful to do so.

The basic difficulty is that there is a certain lack of symmetry in Boole's system. I will make this more precise shortly but it is worth noting first that Boole seemed to be aware of this although not, apparently, of the repercussions it would have. Here is a lengthy quotation that certainly suggests that Boole was keenly aware that his system was constructed to proceed from the simple to the complex and not in the reverse direction:

"I postulate that when the data are not the probabilities of simple events, we must, in order to apply them to the calculation of probability, regard them not as primary, but as derived from some anterior hypothesis, which presents the probabilities of simple events as its system of data, and exhibits our actual data as flowing out of that system, in accordance with those principles which have already been shown to be involved in the very definition of probability.

The ground of this postulate is, that to begin with the simple and proceed to the complex, seems to be, in all questions involving combinations such as we are here concerned with, a necessary procedure of the understanding. The calculation of probability depends upon combinations subject to a peculiar condition, viz., that they shall always present to us a series of cases or hypotheses, to none of which the mind is entitled to attach any preference over any other. We cannot, in endeavoring to ascend from the complex to the simple, secure the maintenance of this condition; but we can do so in descending from the simple to the complex. We have had an illustration of this truth in the reasoning by which we deduced the expression for the probability of the complex event xy from the probabilities of the simple events x and y, supposed to be given. And the methods which have been actually employed in the solution of problems whose immediate data were not the probabilities of simple events, have in fact rested upon the postulate above referred to."[3]

This seems to me to state quite clearly that Boole intentionally constructed his system to proceed from the simple to the complex. Now, the lack of symmetry referred to earlier is that there seems to be no inherent reason why one cannot proceed from the complex to the simple but Boole's system does not provide any method for so doing.

The possibility of so doing surely exists and it is, in fact, completely consonant with the spirit of Boole's method. If one had to state one principle as the basic one exemplified by Boole's method it would certainly be that if a joint probability of various events is not known then it is to be assumed that it has the value it would have if the simple events entering into the joint event were independent. The problem Boole had to overcome in working out the implications of this principle was how to handle the situation that results when some of the events entering into the joint event are themselves joint events.

With this in mind, return to the problem under discussion. Suppose one or more probabilities of simple events are unknown whereas various joint events involving the simple events have known probabilities. Why would it not be equally natural to use the principle that the probabilities of the simple events should be assumed to be those probabilities necessary in order to produce the known joint probabilities under the assumption that the simple events are independent? Clearly, this is merely the obvious extension of the same principle on which Boole based his whole system. in analogy to the problem that Boole had to overcome, we must discover how to accomplish this purpose when one simple event with unknown probability enters into various joint events with known probabilities.

Suppose for three events we are given only p(AB), p(AC), and p(BC) and we want to estimate p(A), p(B), and p(C) as a way to calculate the logical table for the given events. (Note that we will be considering this problem at length in the next Section.) We have a fairly obvious, and quite powerful, approach to this question. This approach is based directly on calculations that would be correct if we were dealing with random, independent events. Suppose this were true for our three events. Then p(AB) = A*B, p(AC) = A*C, and p(BC) = B*C. Therefore, if we are given the probabilities of the three pairs then we can immediately write

$$\frac{p(AB)*p(AC)}{p(BC)} = \frac{A*B*A*C}{B*C} = A*A \quad whence \quad A = \sqrt{\frac{p(AB)*p(AC)}{p(BC)}}$$

and, similarly, we have

$$B = \sqrt{\frac{p(AB)*p(BC)}{p(AC)}} \quad and \quad C = \sqrt{\frac{p(AC)*p(BC)}{p(AB)}}$$

Here we are doing what Boole often does, namely, writing A for p(A) and similarly for B and C, when there is no risk of misunderstanding.

With a little practice, this approach is remarkably powerful. For example, suppose again that we know p(AB), p(AC), and p(BC). Suppose that we simply want an estimate of p(ABC). We can reason as follows. If A, B, and C were random, independent variables then p(AB)*p(AC)*p(BC) = A*B*A*C*B*C = A_2*B_2*C_2 and p(ABC) = A*B*C so p(ABC) = √[p(AB)*p(AC)*p(BC)]. Notice that this argument is equivalent to the immediately preceding argument. Suppose p(AB) = .3, p(AC) = .4, and p(BC) = .25. Then by the equation just deduced we would have p(ABC) = √[.3*.4*.25] = .173205. Using the preceding argument to calculate the probabilities of A, B, and C we find p(A) = .692820, p(B) = .433013, and p(C) = .577350. Using these three values we find p(ABC) = .173205, exactly the same quantity with each argument.

Suppose we know the probabilities of each pair from four variables. Can we estimate A (and B, C, D)? The answer is affirmative and we reason directly from the assumption of random, independent variables. We note that (AB)*(AC)*(AD) = A^3BCD and (BC)*(BD)*(CD) = $B^2C^2D^2$. Therefore, we have

$$A = \sqrt[3]{\frac{(AB)*(AC)*(AD)}{\sqrt{(BC)*(BD)*(CD)}}}$$

Similar expressions can be written for B, C, and D. It is easy to generalize such estimates for any number of variables - under the assumption that we know the probability of each pair of variables. We can attempt to be more perspicuous by writing the formula in prose. For n variables, let

P = product of the probabilities of all pairs involving A;
R = product of the probabilities of all pairs not involving A .

Then, A = (n-1)st root of (P/(n-2)nd root of R). Note that the formula quoted for n = 4 variables is, as it ought to be, a particular case of the general formula.

Perhaps most occasional users of the kind of analysis under discussion here would find the major use of this kind of argument to be the deduction of the additional equations that Boole's method would provide but without having to work through the analysis of the Boolean variables. In fact, many - perhaps all - of the additional equations can be worked out directly from our knowledge that the orthodox theory must apply. Suppose, for example, that with three variables everything is known except p(BC). There is, then, one unknown and one additional equation is needed. Suppose p(A) = p, p(B) = q, p(C) = r, p(AB) = s, p(AC) = t, and p(ABC) = u. Then the logical equations are, using our standard notation:

a = u

$e = s - u$
$b = t - u$
$f = p - s - t + u$
$c = r - t - d$
$g = q - r - s + t + d$
$h = 1 - p - q + s - d$.

We already know a, b, e, and f. The four values we do not know are the four logical classes within A': c, d, g, and h. Following the basic assumption that underlies Boole's system we would expect, then, that BC would act as if it were a random and independent variable within A'. This means that

$$(A'BC) = \frac{(A'B'C)(A'BC')}{(A'B'C')}$$

or c = dg/h, which can equally well be written d = ch/g since "d" is the unknown in our logical equations (of course, it could equally well have been "c"). This is, in fact, the equation that is produced by the Boolean variables in this example.

Suppose that all the information is as in the preceding example except that p(ABC) is also not known. The logical equations are the same as those immediately preceding with "a" substituted on the right-hand side for "u". We would now expect that BC would also act as an independent random variable within A. Therefore, in addition to the equation above we would have the equation or a = be/f. These are the two equations added by Boole's method to the six equations already known in the present example.

So far we have seen that we can use these manipulations based on the assumption of random, independent variables to deduce summary formulas for answers that we otherwise would have to calculate repeatedly and we can also use the same procedures to deduce the logical equations that would be added by Boole's method, thus enabling us to considerably simplify the analysis. In addition to these two uses we can use the same general approach to estimate missing probabilities of simple events from the given information and then apply Boolean analysis to the augmented information. This will always produce Boolean estimates that are in accord with the orthodox theory applied to the augmented information.

Thus, Cayley and Wilbraham maintained that Boole's solution to the Challenge Problem was erroneous. (We have already pointed out that their claim was itself in error since orthodox theory cannot produce estimates for the Challenge Problem as Boole stated it.) The cause of the difficulty is ambiguous since the probability of one of the simple events is not known and, in addition, because of the conditioning of one of the variables d = 0. Our present approach to this kind of difficulty is to suggest that it is usually possible to use the methods we are now considering to estimate, from the data of the example, any unknown probabilities of elementary events. Then we can apply Boole's method to the augmented data. Since there will no longer be any unknown probabilities of elementary events it follows that there will not be any divergence of the Boolean estimates from those of orthodox theory.

We will proceed by constructing an expression for any unknown probability of a simple event. This expression will be in terms of the known probabilities and it will be constructed so that the expression is true for random, independent events. Generally, there will be a variety of such expressions that might be constructed - sometimes an unlimited variety. We will eliminate most such possibilities by imposing two requirements on the expressions that we construct. First, we require that as many of the known probabilities as possible shall be included in the expression. Second, we require that no higher power should be used in the expression beyond the minimum possible. These two requirements are each reasonable and they generally restrict the possible expressions to only one.

Let us assume that p(A) is the unknown probability of a simple event. We have already seen one example of how we might proceed to estimate p(A). Suppose the actual data of the example that is known to us includes p(AB) = p, p(AC) = q, p(AD) = r, p(BC) = s, p(BD) = t, and p(CD) = u. In this example we already know an expression for estimating p(A). Specifically,

$$A = \sqrt[3]{\frac{(AB)(AC)(AD)}{\sqrt{(BC)(BD)(CD)}}}$$

Furthermore, we know that this formula holds true for random, independent events. Note that similar expressions, each based on the same known information, are available for B, C, and D.

Here are some other illustrations of the general procedure. In each case we are assuming that p(A) is unknown. Given B, C, AB, and AC:

$$A = \sqrt{\frac{(AB)(AC)}{(B)(C)}}$$

Given B, C, AB, and BC:

$$A = (AB) * \sqrt{\frac{(C)}{(B)(BC)}}$$

Given B, C, AB, and ABC:

$$A = \sqrt{\frac{(ABC)(AB)}{(B)^2(C)}}$$

Given B, C, AB, AC, and ABC:

$$A = \sqrt[3]{\frac{(ABC)(AB)(AC)}{(B)^2(C)^2}}$$

Given B, C, AB, AC, BC, and ABC:

$$A = \sqrt[3]{\frac{(ABC)(AB)(AC)}{(B)(C)(BC)}}$$

These examples should be sufficient to show that it is generally possible to construct such an expression.

There are quite a few very interesting relationships among the estimates that can be constructed according to this procedure. However, this is a subject that is of only marginal importance for our discussion of Boole's system. The reason for this is that, as previously stated, problems with information missing about the probabilities of one or more of the simple events are not likely to arise very often in real-world applications.

It would now be appropriate to apply this procedure to the Challenge Problem but, unfortunately, this cannot be done because of the other special characteristic involved in this problem. This is the condition that C cannot occur in the absence of both A and B. This means that A'B'C = 0, or d = 0 in the notation we have used for the associated logical table. It is perfectly clear why Boole introduced this condition. Without the condition, Boole's solution to the Challenge Problem would be partly indeterminate. That this is so should be clear from the discussion of the Challenge Problem. In the linear equations based on the logical table it was impossible to separate "d" and "h". In fact, the equation in question was

$$d + h = 1 - p - q + a + e\,.$$

Therefore, given any numerical values for p and q, only the total of d + h can be determined. Since

$$p(C) = a + b + c + d$$

it follows that an estimate of p(C) would contain an unknown part of (d + h) if Boole had not eliminated this possibility by stating that d = 0. This was Boole's motivation for including the requirement that C could not occur in the absence of both A and B - thus requiring that p(A'B'C) = d = 0.

Therefore, we will here again consider an example of the Challenge Problem without the restriction that A'B'C = d = 0. We worked out the Boolean analysis for this variation of the Challenge Problem in Section 27, where we assumed we knew p(A) = p, p(B) = q, p(AC) = r, and p(BC) = s. The major point is that we are **not** assuming that we know p(C). The numerical example we used was p = .4, q = .3, r = .2, and s = .14. In our analysis of this problem type we had five logical equations, including one for "d + h" since we cannot separately estimate these variables. The equation is

$$d + h = 1 - q - s - f + a\,.$$

Since there are seven unknowns (counting d + h as one) we need two additional equations and Boole's method produces

$$a(d + h) = bc$$

and

$$e(d + h) = fg\,.$$

These are somewhat annoying to solve but it is straightforward. We gave the necessary equations in Section 27: a quadratic for "a" plus expressing "f" as a simple fraction involving "a". The reconstructed logical table is (also given in Section 27 but repeated here for convenience)

	A		A′	
	B	B′	B	B′
C	.040837	.159163	.099163	.386497
C′	.045660	.154340	.114340	

The best estimate that can be made, then, for p(C) is

$$P(C) = .299163 + .386497k$$

where the first term is the sum of the three known values for C and where $0 \leq k \leq 1$.

The difficulty with this estimate is not that it is indeterminate but that the argument used to deduce it, Boole's argument, is not in accord with orthodox theory. In order to see this easily, let us write out the estimates for "a" and "f" from the two additional equations in terms of random, independent events. For "a" we have

$$(ABC) = a = \frac{(AB'C)(A'BC)}{(A'B'C + A'B'C')} = \frac{(AB'C)(A'BC)}{(A'B')} = (AC)(BC)$$

and, of course, this equation is not true. Therefore, the estimate of "a" is not in accord with orthodox theory. Similarly, the estimate for "f" is

$$AB'C' = f = \frac{(ABC')(A'B'C + A'B'C')}{(A'BC')} = \frac{(ABC')(A'B')}{(A'BC')} = (AB')$$

and this also is obviously not true for orthodox theory.

Since all of the other estimated logical classes depend on "a" and "f" we are entitled to say that every estimate will be out of accord with the estimates of orthodox theory. To remedy this along the lines being considered in this Section we would first estimate p(C). The obvious estimate is

$$C = \sqrt{\frac{(AC)(BC)}{(A)(B)}}$$

This gives p(C) (= C) = .483046. Now, with this value of C added to the four previously known values we apply Boole's argument as usual. Our known quantities now lead to the following equations:

$$a + b + c + d = .4$$
$$a + c + e + g = .3$$
$$a + b + c + d = .483046$$
$$a + b = .2$$
$$a + c = .14$$
$$a + b + c + d + e + f + g + h = 1 .$$

We easily find the logical equations to be:

b = .2 - a
c = .14 - a
e = .2 - f
g = -.04 + f
d = .143046 + a
h = .356954 - f .

Writing out the Boolean equations (with five variables: A, B, C, t, and v) we find the two added equations are a = bc/d and f = eh/g. Substituting the logical equations and solving gives a = .057965 and f = .138099 and the logical table

	A		A′	
	B	B′	B	B′
C	.057965	.142035	.082035	.201011
C′	.061901	.138099	.098099	.218855

This table can be compared with the preceding one. The major difference is that this new table is in accord with orthodox theory. To verify this, notice that

$$a = (ABC) = \frac{(AB'C)(A'BC)}{(A'B'C)} = (ABC)$$

and, similarly,

$$f = (AB'C') = \frac{(ABC')(A'B'C')}{(A'BC')} = (AB'C')$$

One can also verify that afgd = bceh, within rounding errors. Following this approach we end up with estimates of each logical class in our table. In this exact sense it seems that we are entitled to refer to "a remedy for the error".

We can quickly summarize another example. Suppose that p(A) = p = .63, p(B) = q = .45, p(AC) = r = .21, and p(BC) = s = .18. Again, we assume we do not know p(C). Our logical equations can be written with two unknowns, "a" and "f". The two additional equations from Boole's analysis are a = bc/(d + h) and f = e(d + h)/g. Using the quadratic equation given above for "a" we find a = .079675. Then the formula for "f" gives f = .255575. The logical table is

	A		A′	
	B	B′	B	B′
C	.057965	.142035	.082035	.201011
C′	.061901	.138099	.098099	.218855

Here, again, the best estimate that can be made for p(C) is .310325 + .164100k where $0 \le k \le 1$ and, of course, the estimates violate the Cayley-Wilbraham criterion but, as we have emphasized repeatedly, without knowing P(C) orthodox theory cannot complete the above table so, as always, no comparison is possible between the Boolean analysis and orthodox theory.

We can avoid this problem by estimating p(C) from

$$C = \sqrt{\frac{(AC)(BC)}{(A)(B)}}$$

We find C = .3651. Taking this plus the other four values already known, we apply Boole's method and obtain the logical table

	A		A′	
	B	B′	B	B′
C	.103520	.106480	.076480	.078668
C′	.178624	.241376	.091376	.123476

Here we have obtained, as in the previous example, exact estimates for each logical class and the estimates are all in accord with objective theory, as can be tested by finding that afgd = bceh.

Granted that this seems like a reasonable way to make sure that the Boolean estimates are in accord with those of orthodox theory, let us briefly turn to the first of these two examples and reconsider the possibilities. We assume that we know four of the probabilities: p(A) = p = .4, p(B) = q = .3, p(AC) = r = .2, and p(BC) = s = .14. These, plus our knowledge that the sum of the probabilities in the logical table equals one, mean that we are missing three probabilities that are needed to complete the logical table. Our approach in this Section was to use C = √[(r*s)/p*q] = .483046 to obtain one more probability and then use Boole's method to complete the analysis. This approach assured us that our estimates would be in accord with orthodox theory.

An alternative might be to use the same kind of argument we used to estimate p(C) to estimate the other two missing probabilities. Thus, we could use a similar formula to estimate p(AB). Since we now have C = p(C) we can use

$$AB = \frac{(AC)(BC)}{(C)^2}$$

to find p(AB) = .12. Then we can use

$$(ABC) = \sqrt{(AB)(AC)(BC)}$$

to find p(ABC) = a = .057966. We now have enough information to fill out the logical table:

	A		A′	
	B	B′	B	B′
C	.058	.142	.082	.201
C′	.062	.138	.098	.219

The estimates from this table, too, are in accord with orthodox theory. What, then, is wrong with it? Why can we not take these to be valid estimates? My answer is that if we had no other alternative then these estimates would be enormously better than no estimates. However, I think that the Boolean analysis applied after we estimated only p(C) is considerably preferable because it enforces a consistency among the various estimates of the logical classes that is not built into the immediately preceding table.

There is no doubt that not knowing one or more of the probabilities of simple events will often, perhaps usually, lead to the Boolean analysis diverging from the analysis of orthodox probability theory, when it is possible to apply orthodox theory. It is nonetheless quite difficult to determine exactly how having an unknown probability of a simple event will lead to specific examples of the Cayley-Wilbraham error. As an illustration of the difficulties we can consider again the example we used in Section 13. This example was a seemingly slight variation of the Challenge Problem and, hence, of the example we have just used in this Section. In the Section 13 example we assumed that we knew p(A) = p, p(B) = q, p(AB) = r, and p(BC) = s. For the Challenge Problem we assumed that we knew p(A), p(B), p(AC), and p(BC). Therefore, the only change is that in the Section 13 example we assumed that we knew p(AB) and in the Challenge Problem we assumed that we knew p(AC).Working out the equations for the Section 13 example, we find, including C since our information includes p(BC):

$$c = s - a$$
$$e = r - a$$
$$g = q - s - r + a$$
$$d + h = 1 - p - q + r$$
$$b + f = p - r \quad .$$

We see that now not only will we be unable to separately estimate “d” and “h” but that also we cannot separately estimate “b” and “f”. For the numerical example used in Section 13 we took some actual magazine readership data: p = .161, q = .229, r = .075, and s = .049. Notice from the equations that if we have the estimate of “a” then we can complete the logical table (to the degree that it can be completed). We would therefore like to find an estimator for “a”. The Boolean equations are given in Section 13 and some inspection shows that the numerator of the 2nd & 3rd = that of the 1st & 5th = $uu'vv'AA'B^2$. From the two denominators we get the equation ag = ec. Solving this for “a” we find the desired estimator: a = rs/q. For our numerical example, then, we find a = .016048. The main point of this example is that this estimator is also correct for the orthodox theory. For that theory we would estimate “a” from (AB)(BC)/B = ABC = a and this gives exactly the same estimator as rs/q from the Boolean analysis. Therefore, either analysis gives the same logical table:

	A		A′	
	B	B′	B	B′
C	.016048	.086	.032952	.685
C′	.058951		.121047	

This example shows that it is possible to be ignorant of the probability of one of the simple events (C, here) and still not encounter the Cayley-Wilbraham difficulty. Despite this seeming counter-example, the best summary I can offer is that I am not aware of an instance where the Cayley-Wilbraham problem arises without there being, simultaneously, ignorance of at least one probability of a simple event.

Finally, it is easy to use simple arguments based on the Boolean variables to calculate b, f, d, and h. From (ABC)/AB = C = .213974 we find that b = .086C = .018042. Hence, f = .067598. From C = a + b + c + d = .213974 we now find d = .146572. Hence, h = .538428. This reconstruction was accomplished essentially by using arguments from the Boolean variables to estimate C. Our immediate point is that this would be equally plausible for the orthodox theory as well. This complete logical table would ordinarily be considerably more helpful than the partial one given above.

Section 32. Is a Given Boolean Analysis Unique or Are There Alternatives?

The question of this Section is one to which Boole devoted a lot of thought and, hence, had a strongly held opinion. He argued at some length for his position on at least two occasions. One such argument is given by Boole in the 1862 article which was almost a joint authorship with Cayley (Philosophical Magazine; Series 4; May, 1862; (vol. 123, #155)). There Boole claims that Cayley's solution is not in accord with the "conditions of possible experience" of the data of the Challenge Problem. Boole deduced these conditions and they are quite simple for the Challenge Problem. Specifically, Boole states that it is necessary that

$$p - r + s \geq pq$$

and

$$q - r - s \geq pq$$

as well as the obvious conditions that each of p, q, r, and s should be probabilities. Boole goes on to state that Cayley's solution requires that

$$s \geq rq$$

and

$$r \geq ps\,.$$

Boole says that these two sets of conditions are not equivalent and that Cayley's are more restrictive. Therefore, there are values of the known probabilities for which there are solutions but Cayley's argument will not find them. Boole claims that his method will always produce a solution whenever the known probabilities could actually occur.

It is easy to see that Boole is correct in his statements and in Section 25 there is an example of a set of values of p, q, r, and s that satisfy Boole's "conditions of possible experience" but do not satisfy Cayley's requirements. It is shown in that Section that Cayley's argument does not work with these specific values but that Boole's method gives a solution for this case. Boole apparently considers this to be a major argument against Cayley's solution.

Cayley, on the other hand, cheerfully admits that his solution requires more restrictive conditions than does Boole's. He says that for his argument: "the conditions of consistency ought to be narrower than those in Professor Boole's (solution)". His position here is directly related to his claim that there are really two versions of the Challenge Problem. Cayley maintains that the one Boole stated and that he, Cayley, solved was the "causation version". Cayley believes that Boole solved a different version that Cayley calls the "concomitance version". Section 25 attempts to clarify to some degree what Cayley meant by this distinction. But here it will suffice to note that Cayley argues that the more stringent conditions are both necessary and sufficient for the causation version and that Boole's conditions are only valid for the concomitance version.

Cayley, therefore, is rejecting Boole's argument on the basis that there are two different versions of the Challenge Problem and each one has its own "conditions of possible experience". Boole, on the other hand, is quite clearly making the point that if a proposed solution to a problem does not work for all sets of input data which meet the "conditions of possible experience" then that proposed solution is not correct. Cayley did not offer any argument for his position other than his "intuition" so we cannot simply accept it. We can conclude that if Cayley's two different versions really exist then it may be the case that Cayley is correct. However, with no proof, or even discussion, offered by Cayley we have no basis for accepting his conclusion. Further, we have good reaon to believe that Boole had an integrated approach which would apply equally well to the two different versions, if they exist. In this case, Boole would probably be correct when he would not accept Cayley's approach since it could not accommodate all acceptable sets of input data.

Generally, since neither Cayley nor Boole developed his position far enough we should probably conclude that this example does not resolve the question of this Section: Is a Given Boolean Analysis Unique? Boole's other major example is much more to the point of this question.

We have discussed these examples to some slight degree previously but they definitely deserve some further attention since they play an important role in Boole's thinking and he discussed them quite extensively in the article for which he won the Keith Prize Essay in 1857. These were the two examples that Boole constructed to add cogency to his basic principle that compound events should be treated in the same way as simple events if all that is known is the probability of the compound event. We have previously summarized his examples but as a present reminder we can note that he was arguing that the analysis, by his method, ought to be the same when working an example involving snow, rain, and wind - as an English-speaking person might present it - or whether a person was speaking an imaginary dialect in which the basic terms were "sleet" (the concurrence of rain and snow), "drift" (the concurrence of snow and wind), and "storm" (the concurrence of wind and rain). As an initial comparison, we can see that the English-speaking person would be referring to probabilities of the individual elementary events "rain", "snow", and "wind" whereas the person speaking the dialect would be referring to probabilities of seemingly compound events, defined each in terms of two of the elementary events as used in English. Boole's point seems to be that his analysis ought to give the same conclusion with either of the presentations. We will qualify this conclusion below.

Boole's major point with these two examples is that it should make no difference to the analysis if one works with elementary events as English-speaking persons think of them or with "elementary" events as the dialect-speaker conceives them. Boole's argument is that in these examples and, he claims, generally, the distinction between "elementary" and "compound" is an accident of linguistics. Therefore, he argues, such a distinction cannot and should not affect the results of the analysis by his method. While recognizing Boole's motivation in discussing these examples, we can equally well take other examples of the same kind to illustrate Boole's approach. It is important to remember that Boole's method ought to be applicable to any reasonable presentation of data that gives enough information to permit some estimation. We ought, then, to be able to accomplish Boole's analysis with either formulation.

Therefore, let us consider a media problem - we will use magazines - where we are given data about the effect of pairs of magazines rather than information about the effects of the individual magazines. Suppose, then, that we have information about three pairs of magazines: AB, AC, and BC. Specifically, we know that the pair AB will reach .42 of a the individuals in a specified target population, AC will reach .315 of the individuals in the target population, and BC will reach .27 of the individuals in the target population. More precisely, and in accordance with industry usage, suppose the same advertisement is being run in each of magazines A and B. Then we are given that .42 of the target population will see both of these advertisements. Boole would state that he wished to determine what percentage of the target population will see the advertisement in all three magazines, A, B, and C. As usual, we want to reconstruct the whole resulting logical table.

This is another good example of the very much greater power and ease of our approach as compared to Boole's. Boole gives his analysis in his Section 18 (SLP, pp. 329-330) and I cannot imagine anyone who would not find it a difficult and sometimes seemingly inexplicable algebraic feat.3 Let us make the analysis ourselves. Since we have three compound events we introduce the usual single letters for them. Let us put s = AB, t = BC, and v = AC. Then our information will be of the form, p(s) = p, p(t) = q, and p(v) = r. Let us begin by finding the non-zero entries in the logical table. We have that s′ = A′ + B′, t′ = B′ + C′, and v′ = A′ + C′. By simple and straightforward logical multiplication we find

stv	= ABC	with probability pqr
st′v′	= ABC′	with probability p(1-q)(1-r)
s′tv′	= A′BC	with probability (1-p)q(1-r)
s′t′v	= AB′C	with probability p(1-q)r
s′t′v′	= A′B′C′	with probability (1-p)(1-q)(1-r).

All of the other entries in the logical table are zero. The table looks like this:

	s		s′	
	t	t′	t	t′
v	ABC			AB′C
v′		ABC′	A′BC	A′B′C′

Using our letters for this size table we now know that b = c = e = 0. Therefore, our information can be summarized by the equations

$$a + f = p$$
$$a + g = q$$
$$a + d = r$$
$$a + d + f + g + h = 1\,.$$

We can write all of these equations conveniently in terms of one unknown, "a":

$f = p - a$
$g = q - a$
$d = r - a$
$h = 1 - p - q - r + 2a$.

Then, as usual, we write the Boolean equations with denominators from the logical table. We have:

$$\frac{stv}{a} = \frac{s't'v}{d} = \frac{st'v'}{f} = \frac{s'tv'}{g} = \frac{s't'v'}{h} = V$$

Here we need only one equation so that we can solve for "a". By inspection we discover that the numerator of the product of the 2nd * 3rd * 4th equals the numerator of the product of the 1st * (5th)2. The numerator in each case is $stv(s')^2(t')^2(v')^2$. Therefore, we can equate the denominators to get $ah^2 = dfg$. Substituting our expressions for these logical classes gives us the desired equation for "a"

$$a(1 - p - q - r + 2a)^2 = (r - a)(q - a)(p - a) .$$

This is a cubic equation for "a" and we can easily obtain the desired root numerically. For our example we have p = .42, q = .27, and r = .315. To a sufficient approximation we find a = .12701. This, of course, would be the answer to Boole's question: a proportion .12701 of the target population would see all three advertisements. Using our equations, we can find the proportion in each of the possible logical classes:

ABC	=	a	=	.127
ABC′	=	f	=	.293
A′BC	=	g	=	.143
AB′C	=	d	=	.188
A′B′C′	=	h	=	.249 .

It is worth mentioning that we could equally well determine s, t, v, and V if we so desire. Thus, using the 2nd and 5th we get v = .43017; from the 3rd and 5th we find s = .54056; and from the 4th and 5th we get t = .36476; whence, V = .66781. (These values were obtained using a = .12701.)

Suppose, instead, that we want to construct our analysis in terms of the elementary events A, B, and C. We know p(AB) = .42 = p, p(AC) = .315 = r, and p(BC) = .27 = q. Suppose we let the unknown p(A) = α, p(B) = β, and p(C) = γ. Then we can immediately set up three equations in three unknowns:

$\alpha\beta = p$
$\alpha\gamma = r$
$\beta\gamma = q$.

These equations are easy to solve and we find:

$$\alpha = \sqrt{\frac{pr}{q}} \quad and \quad \beta = \sqrt{\frac{pq}{r}} \quad and \quad \gamma = \sqrt{\frac{qr}{p}}$$

For our numerical example we find α = p(A) = .7, β = p(B) = .6, and γ = p(C) = .45. We do not need anything further by way of Boole's method because we know that when the elementary events have known probabilities and nothing else is known then the logical table is exactly the same as the orthodox theory would give for random, independent variables. Thus, this logical table would be

	A		A′	
	B	B′	B	B′
C	.189	.126	.081	.054
C′	.231	.154	.099	.066

This logical table is, in fact, quite totally different from the previous one. That table, written in terms of A, B, and C is

	A		A′	
	B	B′	B	B′
C	.127	.188	.143	
C′	.293			.249

The differences between the two logical tables highlights the fact that it is not reasonable to assume that Boole did not notice this difference. Using logical multiplication on our variables s, t, and v we discovered that it was necessary to have none of the three letters negative, or two of the three letters negative, or all three of them negative. Any case with just one of the three negative was logically impossible. These facts translate into equivalent statements in terms of the logical table for A, B, and C. It is logically impossible to have exactly two of these three events negative. That is why in the immediately preceding logical table there are no entries for AB'C', A'B'C, or A'BC'. On the other hand, obviously, any one of the eight possibilities can occur when we use orthodox theory and assume we know the probabilities of the elementary events, A, B, and C.

The two analyses could not be more different. The simplest statement of the reason for the difference is that basically the two formulations are incommensurable. A trial when the elementary events are A, B, C occurs when each of the three letters or its negation occurs. A trial when the elementary events are s, t, v occurs when each of s, t, v or its negation occurs. But each of these elementary events involves two of the elementary events A, B, and C, either in the sense that two of them occur or in the sense that one of two is negated. It is easy to see that it is logically impossible to have two of s, t, and v occur positively and the third one be negated. It is these differences in the way the elementary events interact in the two formulations that result in the major differences in the analyses.

It is, therefore, dubious to assume that Boole did not recognize such a major difference in the analyses. In fact, his writing here is not crystal clear. He discusses two different formulations of the example assuming the dialect and he definitely shows that these two different formulations lead to the same conclusions. In addition, he briefly discusses a third formulation based on the English language. For this one he shows that the "conditions of possible experience" are different from the other two formulations. He does not advert to the major differences between this analysis and the other one that we have shown here. I conclude that there is no basis for accusing Boole of an error here - particularly because the question involved does not have any effects on our subsequent development of the underlying idea.

However, this granted, it remains true that Boole is certainly wrong when he argues that it is possible to choose the “better” model based on the “conditions of possible experience”. The correct model for this example depends entirely on the details of the procedure that will be followed in simulating an instance of the process. If A, B, and C are being picked randomly and independently then the correct logical table is the one based on A, B, and C and with no entries in the table being zero. If the random variables s, t, and v are being used - each one involving two of A, B, and C - then the logical table should be based on s, t, and v. Instead, the equivalent table based on A, B, and C can be used and this table will have three zero entries.

Perhaps, or probably, if Boole were alive to respond to this argument he would object that the persons speaking the dialect - equivalent to using s, t, and v - could not logically use a table based on A, B, and C since that would violate the initial assumption that the dialect did not have the simple events of English: snow, rain, and wind. However, in Section 12 it was already argued that it is, in fact, impossible for the speakers of the dialect not to be aware that “snow”, “rain”, and “wind” are the simple events and that “sleet”, “drift”, and “storm” are compound events. If they were not so aware there would be no way in which they could define the usage of the three terms: “sleet”, “drift”, and “storm”.

It appears that we can escape the conclusion that it is not possible to choose the “better” model based on the “conditions of possible experience”. This can be achieved by explicitly recognizing that it is necessary to state the input information for a given problem **and** the exact characterization of the procedure that would have to be followed in order to simulate an instance of the process in question. However, this is not generally Boole’s approach. In almost all cases he assumes that the problem is sufficiently defined simply by giving the input information. This is definitely not correct. We will see that an analogical difficulty arises with regard to Boole’s treatment of “conditioning”.

We can summarize the underlying relationships which lead to an equation that determines “a”. It is convenient to have suitable notation. For four simple events let us represent the product of all possible joint events having exactly one of the variables negated as

$$\pi(4,1) = (ABCD')(ABC'D)(AB'CD)(A'BCD)$$

and the product of all possible joint events having exactly two of the variables negated as

$$\pi(4,2) = ABC'D')(AB'CD')(A'BCD')(AB'C'D)(A'BC'D)(A'B'CD)$$

and so forth. Then the probability of a = ABCD is given by

$$(ABCD) = \frac{\pi(4,1)\pi(4,3)}{\pi(4,2)\pi(4,4)}$$

Substituting for the logical variables gives

$$a = \frac{bcehilno}{dfgjkmp}$$

using the obvious extension of the previous notation for the logical variables. Substituting from the known logical equations gives an 8th-degree equation for "a".

For five variables the added equation is

$$(ABCDE) = \frac{\pi(5,1)\pi(5,3)\pi(5,5)}{\pi(5,2)\pi(5,4)}$$

which gives a 16th-degree equation for "a". The general equation for n variables is

$$(ABCD...N) = \frac{\pi(n,1)\pi(n,3)\pi(n,5)...}{\pi(n,2)\pi(n,4)...}$$

which will produce an equation for "a" that is of the 2^{n-1}th-degree. Obviously, the complexity of the equation increases sharply with the number of simple events. This is simply a reflection of the fact that the size of the logical table increases rapidly with the number of simple events. For example, with 10 simple events the logical table will have 2^{10} = 1024 cells and the equation for "a" will be of the 512th degree. While it is obviously possible to solve such equations on computer it is equally obvious that the usefulness of such a solution is very marginal. In Chapter 5 we will see that there is an alternative approach when n is large and that this approach leads to a far more useful output.

It is surely worth illustrating the difficulties that arise as the size of the underlying logical table increases. Suppose, then, that we know the singles, duplications, and triplications for four magazines and that the information is:

p(A) = .383	p(AB) = .185	p(ABC) = .097
p(B) = .334	p(AC) = .173	p(ABD) = .090
p(C) = .346	p(AD) = .154	p(ACD) = .085
p(D) = .310	p(BC) = .158	p(BCD) = .079
	p(BD) = .151	
	p(CD) = .136	

Since we know the triplicates we can use this information in writing our equation. Our logical table will have 2^4 = 16 cells and the small letters will run from "a" in the upper-left cell to "p" in the lower-right cell. Thus, in the present case there will be an 8th-degree equation to determine Boole's estimate of "a" = ABCD. The equations for the small-letter variables can all be written with the unknown quantity being "a". Thus, these equations are:

b = .085 - a	c = .079 - a	d = -.028 + a	e = .097 - a
f = -.009 + a	g = -.018 + a	h = .140 - a	i = .090 - a
j = -.021 + a	k = -.018 + a	l = .123 - a	m = -.002 + a
n = .143 - a	o = .106 - a	p = .233 + a	

As given above, the desired 8th-degree equation to determine "a" is simply adfgjkmp = bcehilno, written in terms of the above logical equations:

$$a(-.028+a)(-.009+a)(-.018+a)(-.021+a)(-.018+a)(-.002+a)(.233+a) =$$

$$(.085-a)(.079-a)(.097-a)(.140-a)(.090-a)(.123-a)(.143-a)(.106-a)$$

We want to find the value of "a" that gives an equality. Let the left-hand side be f[a] and the right-hand side be g[a]. Clearly, the values of f[a] and g[a] will be very small for the equalizing value of "a". This makes it somewhat troublesome to solve the equality but it is fairly straightforward using Mathematika. I find that for a = .05345 we have f[.05345] = 3.634031164 E-11 and g[.05345] = 3.634697948 E-11. This is a completely adequate precision and we can use our discovered value of "a" and the logical equations to reconstruct the table:

		A		A′	
		B	B′	B	B′
C	D	.053	.032	.026	.025
	D′	.044	.044	.035	.087
C′	D	.037	.032	.035	.070
	D′	.051	.090	.053	.286

We have accomplished our purpose but was it worth doing?

It seems quite doubtful that it was worth doing. Notice that many of the cells have very similar numerical values. Thus, ten of the cells have numerical values between .032 and .053. Inspection of the logical equations suggests that the differences between cells may very well have been due more to random fluctuations in the input data than to actual differences between the cells. The thrust of this point of view is enhanced when it is recalled how the resulting table is intended to be used. Ultimately, the hope is that such tables will permit the relevant decision makers to estimate values that might be achieved by alternative selections of magazines (and other media as well) and thus to optimize the return from an investment in advertising. This seems to demand almost superhuman abilities in making intuitive estimates of the differences between magazines due to different editorial climates, for example. Suppose that an advertiser has decided to use A and B and is trying to decide whether it would be better to add C or to add D to these two. According to our table C without D will reach .044 and D without C will reach .037. A small improvement in compatibility between D and the target audience would make D a better investment than C. Who is able to make such accurate judgments?

The question that we are considering becomes more and more important at a relatively rapid rate as the number of magazines (or media in general) increases. Let us suppose that the result we here illustrated for four magazines is adequate. It will certainly **not** be adequate for any larger number of magazines. We need a much simpler argument and a different kind of presentation of the data. We have such an argument and presentation and it will be discussed more lengthily in Chapter 5. Here it is sufficient to state that the goal becomes one of estimating the distribution of the number of exposures achieved by different groups of magazines. In order to do this for our data we use Jaynes' Principle (to be discussed in Chapter 5) to estimate "a" of our logical table and then use standard formulas (given in Chapter 5) to estimate the exposure distribution. For our example we find:

Exposures	Probability
0	.288
1	.294
2	.231
3	.133
4	.054

The estimate of "a" on which this table is based is a = .054 or, more accurately, a = .0545585. It is generally thought, by persons with long experience in the advertising industry, that it is very much easier to estimate the value of an additional exposure than it is to estimate increments of value among the cells in a logical table. This does **not** mean that it is easy. The operative word is "easier".

One other point can be made from the data of our present example. Suppose that we do not know the triplicates. In this case we could write a single equation with five unknowns: one each for the unknown triplicates and one for the unknown a = ABCD. We would then have an 8th-degree equation with five unknowns and this would be expected to be difficult to solve. In most cases we would want to convert this to an application of Jaynes' Principle. The present point is that this can be avoided by using a procedure that is easy to execute on a computer and which remains well within the bounds of hand calculation (using a calculator) for up to five magazines. The procedure is this. Take only the data for magazines A, B, and C and use the equation given earlier in this Section to predict (ABC); then do the same thing for magazines A, B, and D; and similarly for the other two combinations of three magazines at a time. This gives estimates of the four triplicates. Use these estimates with the known data and the equation, illustrated with this example, to estimate p(ABCD). It can be shown that this will give the same estimates as if all five unknowns were estimated simultaneously. This is a very major simplification and the underlying idea is often available.

Section 33. Boole's Conditioned Variables.

33-A. Boole's Theory and Donkin's Principle.

Repeatedly in his examples Boole introduces equations to reflect the restrictions on those random variables that Boole calls "conditioned". These are combinations of random events that either cannot occur or must occur. The restriction that C cannot occur in the absence of both A and B in the Challenge Problem is a typical example. Boole explicitly states that this is his procedure: "Should the data, besides informing us of the probabilities of events, further assign among them any explicit connexion, such connexion must be logically expressed, and the equation or equations thus formed be introduced into the general system."[4] It is evident that he believes that this procedure is correct and in strict accord with his general theory. Now, while this is often true, it is not always true. In particular, it is not usually true in precisely the circumstances where Boole apparently thought it would be true. Obviously, some clarification is needed. This is particularly the case since Boole himself made his most glaring mistake because of confusion about this question.

The major reason why Boole accepted the procedure as quoted in the previous paragraph was the high regard he held for Donkin's principle plus the fact that the essential step in Boole's general argument seems to rest four-square on that principle. Adding equations describing conditioned events seems simply to be doing what his system generally does anyhow. These two points need some demonstration.

That Boole had high regard for Donkin's principle is evident from his comments about it. When Boole is enumerating the principles on which the theory of probability is based, he concludes with:

"To the above principles we may add another, which has been stated in the following terms by the Savilian Professor of Astronomy in the University of Oxford.

"Principle 8. If there be any number of mutually exclusive hypotheses, H_1, H_2, H_3, ... of which the probabilities relative to a particular state of knowledge are P_1, P_2, P_3, ... and if new information be given which changes the probabilities of some of them, suppose of h_{m+1} and all that follow, without having otherwise any reference to the rest; then the probabilities of these latter have the same ratios to one another, after the new information, that they had before, that is,

$$P_1' : p_2' : p_3' : \ldots : p_m' = P_1 : P_2 : P_3 : \ldots : p_m$$

where the accented letters denote the values after the new information has been acquired.

This principle is apparently of a more fundamental character than the most of those before enumerated, and perhaps it might, as has been suggested by Professor Donkin, be regarded as axiomatic. It seems indeed to be founded in the very definition of the measure of probability, as `the ratio of the number of cases favorable to an event to the total number of cases favourable or contrary, and all equally possible'."[5]

Boole goes on to suggest that this principle appears to be inseparably involved in the very definition of probability. It is surely true that Boole held this principle in particularly high regard.

Donkin's principle was widely recognized as being a very important formulation of a basic postulate of probability theory. Let us attempt to specify more exactly when it applies and when it does not. The area of application is given quite precisely in Donkin's statement of the principle. The key qualification to the applicability of the principle is the clause, "without having otherwise any reference to the rest". in other words, for the principle to apply it is necessary that the fact that we are eliminating some hypotheses from consideration should not have any effect on the likelihoods of the hypotheses still remaining. Thus, some group of people is categorized according to type of work: lawyers = .09, doctors = .13, other professional = .19, blue collar = .31, and so forth. Now, if for some reason we want to consider only the various proportions among the professionals we would immediately conclude that, for example, doctors were .13/(.09 + .13 + .19) = .317. This would be an application of Donkin's principle. Perhaps the commonest use is the argument routinely made in calculating conditional probabilities. For example, suppose we are given the joint probability table:

	A	A'	
B	.12	.28	.4
B'	.08	.52	.6
	.2	.8	

When we calculate a conditional probability, such as

P(A|B) = .12/.4 = .3,

we are using Donkin's principle when we ignore the row representing B' and confine our attention to the row representing B. Such uses of Donkin's principle are unexceptionable. Indeed, they are so routine and obvious that it is easy to conclude that the principle has a considerably greater range of applicability than is in fact the case. Perhaps Boole slipped into this mistake.

There is no question that the correct use of Donkin's principle plays a key role in Boole's system. Boole's entire approach depends on the quantity he designates as V, which equals the sum of all of the possible combinations of the Boolean variables. The key word here is "possible". We can take Boole's solution of the Challenge Problem as an illustration of the present point. There are four Boolean variables: A, B, s = AC, and t = BC. We did not bother to obtain the numerical values of these variables previously because they are not needed in order to find the desired answer or for the reconstruction of the logical table. However, they can be found easily enough:

A = .450456377

B = .250253543

s = .766508960

t = .597583148 .

Clearly, if we treat these simply as arbitrary symbols there are sixteen terms that involve these four and/or their negations. Each one of these terms has the corresponding probability associated with it. Thus,

P(A'B'st) = (.549543623)(.749746457)(.765508960)(.597583148)

= .188480 .

But, of course, this particular term is logically impossible since

A'B'ACBC = 0 .

Now, Boole's V is precisely the sum of the probabilities of the seven combinations of the four Boolean variables that are, in fact, possible. Specifically,

P(stAB)	=	.051635520
P(st'AB')	=	.104174292
P(s'tA'B)	=	.019188939
P(s't'AB)	=	.010592034
P(s't'AB')	=	.031733176
P(s't'A'B)	=	.012921972
P(s't'A'B')	=	.038713548
V	=	.268959480

Boole's argument now involves a perfectly legitimate application of Donkin's principle. Since the only logical combinations that are, or could be, observed are those listed it follows that their observed probabilities must be the calculated quantities divided by V. Thus,

P(ABC) = P(stAB)/V = .051635520/.268959480
= .191983 .

This argument is legitimate because the knowledge that the combinations in V' did not occur (since they are logically impossible) clearly has no effect on the relative likelihoods of occurrence of the combinations in V. This is the only requirement needed in order to apply Donkin's principle.

Boole was obviously completely familiar with this application of the principle at the very heart of his theory. Apparently he believed that there was a considerably greater correspondence between the areas of application of Donkin's principle and his own theory than was actually the case. This led Boole to make a particularly glaring mistake that deserves a separate Section.

33-B. Boole's Note.

Boole's first example dealing with probabilities involves drawing balls from an urn and noting only whether the ball is marble or white or both, with no record being kept of the number of times when it was neither the one nor the other. This example was already mentioned in Section 13 because it is the example that Wilbraham specifically objected to and it was, in effect, withdrawn by Boole in the Note added (presumably while his book was in press) before the title page of "An Investigation of the Laws of Thought".

Under the stated circumstances, suppose the probability that the ball was marble has been observed to be q and the probability that the ball was white has been observed to be p. Boole proposes the problem: "from these data alone let it be required to find the probability that in the next drawing, without reference at all to the conditions above mentioned, a white ball will be drawn; also the probability that a marble ball will be drawn."[6]

In his original text - that is, without the emendation of the Note - Boole gives the solution:

$$p' = \frac{p + q - 1}{q} \quad and \quad q' = \frac{p + q - 1}{p}$$

After giving this solution, Boole makes the following remarkable statement: "It is seen that p' and q' are respectively proportional to p and q, as by Professor Donkin's principle they ought to be. The solution of this class of problems might indeed, by a direct application of that principle, be obtained."[7] I call this statement remarkable because in customary usage one simply would not ordinarily say that p' and q' are proportional to p and q. The statement is dubious from any point of view. The solution that Boole offers in the Note, p' = cp and q' = cq, is in accord with Donkin's principle but the original solution, given above, is completely out of accord with it. How could a person of Boole's mathematical abilities make such a unusual statement concerning such a simple matter? I believe that this can be most easily understood if it is recognized that Boole on the one hand accepted Donkin's principle as being generally true. On the other hand, he also believed that his own method was always true. Therefore, whenever it was possible to apply both, Boole's natural assumption was that they must agree. I suspect that his belief in this was so basic that he didn't even notice that they did not agree in the cited example. This, of course, is speculation but I believe that it is reinforced by the uncritical acceptance by Boole of Donkin's principle as justification for his treatment of "conditioned" variables.

The solution given by Boole in the text, quoted above, is perfectly correct and in strict accord with his system. I used the same example in a different guise in Section 13 (a company with two different files of customers who obtained two different merchandise premiums). It will be seen that the solution given in Section 13 is the same as Boole's. Therefore, it is even more remarkable that in the Note Boole disavowed his solution. He replaces it with a very general "solution" that does not provide a specific answer at all. He does this, according to his own statement, in order to make the solution in accord with Donkin's principle. This, perhaps more than any other instance, serves to demonstrate the importance that Boole ascribed to Donkin's principle. The general solution that Boole offers in his Note is

$$p' = kp \quad \text{and} \quad q' = kq .$$

But we know that Boole's method gives a unique value of k and when we find it we get exactly the same answer as Boole's text solution.

So now it should be clear what the Note incorrectly did. in effect, it gave up Boole's whole general system as applied to this example in favor of a solution that is in overt accord with Donkin's principle. It should be clear that it is not at all possible, as Boole thought, to deduce the solution of this example by using Donkin's principle. Indeed, Boole of all persons should have been the most immediately familiar with this fact. As he carefully noted, Donkin's principle is only part of standard probability theory. in this example, as in many of Boole's examples, the question being asked cannot be answered by standard probability theory. in fact, the present example is just like the Challenge Problem in this regard.

Boole's Note surely suggests that Boole relied excessively on Donkin's principle. This may account for his giving insufficient attention to the circumstances under which "conditioned" variables can be handled by adding an appropriate equation to the initial system of equations.

33-C. Donkin's Principle and Conditioned Variables.

The completely correct use of Donkin's principle illustrated in Section 33-A is basic to Boole's whole system. From this use it seems to be a short and easy step to defining V, if necessary, to exclude other combinations of the Boolean variables when some of these are "conditioned". This is the step that Boole took. However, it is not necessarily correct. Stating that some logically possible combination of variables cannot occur is not at all the same thing as having a combination actually be logically impossible. Such a statement often has an effect on the likelihood of the possibilities remaining, in contradiction with the requirement for the applicability of Donkin's principle. Boole apparently gave little thought to this question of whether Donkin's principle remained applicable when equations representing "conditioned" variables are included in the system of equations.

In any case, the question that needs to be raised is: under what circumstances will Boole's method produce the correct, and expected, answer to a probability problem when an equation representing a conditioned variable has been added to the system of equations describing the known data? If it does produce the correct answer it may or may not be due to the operation of Donkin's principle. The important question, obviously, is whether it is correct, not whether Donkin's principle applies or not. In fact, we shall see that Donkin's principle has very little to do with the question whether or not this procedure is correct even though this was Boole's own argument for the procedure.

The basic difficulty is that there are always at least two possible interpretations of a condition such as p(A'B'C) = 0 in the Challenge Problem. Often there are more than two. But two alternatives is sufficient to demonstrate that no system, including Boole's, that has one automatic procedure can possibly be correct for both alternatives.

An example may be useful. Let us consider the Challenge Problem except that we will assume we know p(C). Specifically, suppose p(A) = p = .63, p(B) = q = .45, p(C) = r = .3, p(AC) = s = .21, and p(BC) = t = .18. Now, without the additional restriction that p(A'B'C) = d = 0, the logical equations are

$$b = s - a$$

$$c = t - a$$

$$f = p - s - e$$

$$g = q - t - e$$

$$d = r - s - t + a$$

$$h = 1 - p - q - r + s + t + e \quad .$$

There are two unknowns, "a" and "e". The two additional equations added by Boole's method are

$$a = st/r$$

and

$$e = (p - s)(q - t)/(1 - r) \quad .$$

The reconstructed logical table is:

	A		A′	
	B	B′	B	B′
C	.126	.084	.054	.036
C′	.162	.258	.108	.172

Now, suppose we add the restriction that d = 0. From the logical equations, this means that

$$d = 0 = r - s - t + a$$

whence

$$a = s + t - r \quad .$$

Obviously, the information that d = 0 has acted to eliminate one of the unknowns. More directly, the added equation that involved "d" has been eliminated. The other equation, however, has remained the same. The reconstructed logical table now is:

	A		A′	
	B	B′	B	B′
C	.09	.12	.09	0
C′	.162	.258	.108	.172

It will be seen that the row representing C' is the same in this table as in the one without the restriction that d = 0. This was necessarily true because the additional equation affecting this row (e = (p - s)(q - t)/(1 - r)) remained the same.

Each of these two tables resulted from a straightforward application of Boole's system. But suppose that the first of these two tables holds and we are told that for some reason A'B'C was simply not observed. Then Donkin's principle surely holds and we would find the new logical table by setting "d" = 0 and dividing each of the other seven entries in the table by 1 - d = .964. This gives a completely different table than our second one:

	A		A′	
	B	B′	B	B′
C	.131	.087	.056	0
C′	.168	.268	.112	.178

Therefore, we can see immediately that Boole's method does **not** produce the answers required when Donkin's principle applies.

To put the matter in reverse, if we were given the second table then the application of Boole's method would produce the first table. The "application of Boole's method" means that we would set

$$a = st/r = .126 \qquad d = r - s - t + a = .036$$

$$b = s - a = .084 \qquad c = t - a = .054\ .$$

But if the assumption underlying Donkin's principle was the reason for d = 0 in the observed table then we would want a completely different table that would not be produced by Boole's method. The actual original (unknown) value of "d" cannot be determined but the structure of the reconstructed table would be known: the sum of the seven non-zero values multiplied by any (1 - d) and with the zero entry replaced by "d" will sum to one and would be an allowable reconstruction of the table for any 0 < d < 1. Further, there are various other possibilities, any one of which might be true in a specific actual case. To continue the present example and assuming that we are given the second table, perhaps d = 0 refers only to the row representing C. Thus, in terms of Boole's method this row might be handled by itself. In terms of Donkin's principle, only this one row would be affected. Or it may be that d = 0 refers to the whole table and we have the two alternatives first discussed.

Boole apparently thought that adding an equation for a conditioned variable had a unique meaning and that his method captured this meaning. There is no doubt that his method does capture a possible meaning but it does not follow that it is always the correct meaning.

Generally, the problem is that if some event is "conditioned" it is not ordinarily clear exactly what this means. At one extreme it could mean simply that some combination of events was not recorded, thus leading to Donkin's principle. At the other extreme, it might mean that some combination of events was physically impossible. This latter case is the one that Boole's method assumes. There are various intermediate possibilities in many instances. This granted, then it is a fair summary to state that if the actual effect of "conditioning" on all of the variables can be exactly stated, then, and only then, it is possible to incorporate the appropriate equations in the Boolean system. It is certainly not true that it will always be correct simply to add a corresponding single equation to the system of equations, as Boole apparently thought.

Section 34. Boole's Theory of Probability

In Chapter 1 in my discussion of the Challenge Problem I said that George Boole had bad luck in his choice of examples. This is surely true. There are only two difficulties with the Boolean theory of probabilities that he did not himself resolve. That the Challenge Problem should include both of them was extremely bad luck. It is my opinion that if Boole had selected a different problem, without the two difficulties, there might well have been a completely different response to his theory.

In this Chapter I have attempted to show that the difficulty that most persons thought was the coup de grâce for Boole's theory, the Cayley-Wilbraham criticism, in fact is irrelevant to Boole's theory. Furthermore, clear quotations from Boole demonstrate beyond any doubt that he was thoroughly aware that this criticism was irrelevant to his theory. The reason the criticism is irrelevant is that the criticism assumes that there is an "orthodox probability theory" answer to the problem at issue and that Boole's solution diverges from that one. But this never happens. Whenever the orthodox probability theory can produce an answer then Boole's theory will produce the same answer. But there are very many kinds of probability problems for which the orthodox theory cannot produce an answer and Boole's theory can. It is on these that critics of Boole's theory need to concentrate their attention. We have shown in this Chapter that the supposed divergence of Boole's solutions from orthodox ones arises primarily when at least one of the simple events involved does not have a known probability. In the title of this Chapter, "An Extension" refers to our suggestion that this whole question can be avoided by using appropriately constructed Boolean variable to estimate any such missing simple probabilities. This approach is discussed in Section 31.

In the title of this Chapter, "A Correction", refers to our discussion of Boole's treatment of conditional variables. Here there is definitely an error in Boole's approach and I have not previously seen this question discussed. It turns out that there is considerably more ambiguity in the idea of "conditioning" than Boole suspected. Section 33 undertakes to show that if a careful statement of exactly what is meant by the conditioning is constructed then the conditioning can be handled by adding an appropriate equation to the problem description. Boole apparently did not realize that there was considerable ambiguity in the meaning of some specific conditioning and, hence, he thought that one kind of equation was always adequate. This is definitely incorrect.

But above any such criticisms, it needs to be emphasized that innumerable Boolean problems - the vast majority - do not involve either of these questions and can be analyzed exactly as Boole proposed.

The great majority of real-world applications would involve nothing but straightforward Boolean analyses of exactly the kind George Boole gave us.

In Chapter I in my discussion of the Challenge Problem I said that George Boole had bad luck in his choice. This is surely true. There are only two difficulties with the Boolean theory of probabilities that he did not himself resolve. That the Challenge Problem should include both of them is extremely bad luck. It is my opinion that if Boole had selected a different problem, without the two difficulties, there might well have been a completely different response to his theory.

He also had bad luck in having Cayley find a way to solve his Challenge Problem with a remarkably innovative use of orthodox theory. Then, in addition he had the misfortune of having Cayley state to the whole audience of interested persons that he, Cayley, did not understand Boole's argument. This was in spite of Boole's repeated claims that his method was based entirely on the observed data and that no hypotheses were introduced by it. It is one thing for a person of modest attainments to claim lack of understanding but it is quite a different thing for one of the most famous of contemporary English mathematicians to make such a claim.

Further, Boole had bad luck in the fact that the applications of his theory, as he presented it, in most cases led to phenomenal difficulties for the pencil-and-paper calculations to which everyone was then confined. I do not think that it is far from the truth for me to doubt whether anyone in the past 100 years has undertaken a Boolean analysis of probabilistic data, meaning by this the use of Boole's own methods. It is surely true that the major published books during this period have not done so. (These last two sentences are with the extremely important exception of Hailperin's fine book.)

Finally, Boole had bad luck in the fact, already mentioned earlier, that the advent of Frege and Russell & Whitehead put an end to the exploration of the interrelationships between the algebra of logic and probability theory. The whole potential development of the ideas that he had presented was aborted. There is a whole area of books and articles that ought to have refined Boole's ideas but which simply never happened. That is quite a bit worse than just "bad luck". It is more nearly a tragedy.

Boole was a kindly and modest man. His life was foreshortened - in the beginning by poverty, in the end by his untimely death. I believe that the great conception he had and presented to us in his "An Investigation of the Laws of Thought" was equalled in human intellectual history only by those of Aristotle and Leibniz. It seems to me that he deserved far better from his posterity than he received. He succeeded fully in his goal and no one seems to know it. If anyone happens to read this book and as a result shares my own enthusiasm for the remarkable intellectual structure that George Boole created then this author's modest efforts will be richly rewarded.

CHAPTER V

SOME APPLICATIONS OF BOOLE'S THEORY

Section 35. Guide to This Chapter

In this Chapter I will offer evidence in support of my opinion that Boole's theory of probability has some important applications to real-world problems. While I do not intend to argue that Boole's theory of probability is the only way in which those problems can be attacked, I do want to emphasize that Boole's theory has a natural aptitude for one whole class of problems. These problems arise with great frequency in advertising but they also arise very often in most scientific studies that involve contingency tables.

In order that the reader may understand the examples dealing with media data (primarily magazine data) I will devote Section 36 to a discussion of the Boolean logic that underlies media data. There are four major ways in which the same information may be presented for media data. Three of these ways are complete in the sense that each of them gives all the information about the group of media included. This means that it is possible, and often important, to transform the data from one of the three ways to another which is more convenient for the purposes of the advertiser. The fourth mode of presentation does not provide complete information but it is, nonetheless, from the advertiser's point of view perhaps the most important presentation of the information about the media. We cannot devote very much space to the details of these different presentations but we need at least to make the reader familiar with the basic kinds of presentation of data that underlie the applications of Boolean methods.

In Section 37 I will summarize some of the non-Boolean methods of analysis of media data that are available to attempt to help the advertiser achieve some kind of optimal selection of media. It is my impression - but I am not certain - that some of these methods are quite useful and are, perhaps, not as widely known as they might be. There will be four sub-sections in Section 37, each one dealing with a different kind of approach. Sub-section A will briefly present linear programming as the major method for obtaining bounds on desired probabilities. Sub-section B will summarize the argument that underlies the use of Jaynes' Principle, a very widely used approach in dealing with decision problems with partial information. Sub-section C will present an approach based on the idea that it is often possible, and reasonable, to change the stated reaches of magazines so that the individual magazines become more nearly random and independent variables. Finally, sub-section D will discuss an interesting relationship between standard regression analysis and the logical tables that we have used so often.

In Section 38, then, we will consider how Boole's theory of probability can be applied to problems involving media data. We will find that the potential area of applications is very large. Many of the kinds of breakdowns that decision makers in advertising want to have are almost "made to order" for Boolean analyses and there is no completely satisfactory alternative known to me for constructing these breakdowns. We can use the general formula deduced in Chapter IV to quickly calculate the algebraic equation that has the key quantity to be estimated as its root. With this root we can easily determine the probabilities of each logical class for the magazines under consideration.

The major problem considered in Section 38 can be stated this way. Suppose the advertiser only has data for sets of three magazines. This is quite a common situation. Perhaps the advertiser has purchased data for 50 different magazines but all that is given is the information about each of the 1960 sets of three magazines. Suppose the advertiser is considering using seven magazines in his or her campaign. How can the advertiser use the available data to estimate the probabilities of each of the 128 logical classes that will result from seven magazines? This would be a standard kind of usage of the general formulas developed in Chapter IV.

However, there is an underlying question that arises for magazine data but also for virtually any contingency table from any field. We briefly mentioned this general question in our discussion in Chapter IV. In the previous paragraph we suggested an advertiser who was considering using seven magazines and only had data by threes. Let us change the example to an advertiser who is hoping to use five magazines and who has the good fortune to have magazine data for all sets of 5 magazines. (For a list of 50 different magazines this would require 2,118,760 different sets of five magazines each - an enormous, and very expensive data base.) As we emphasized in Chapter 4, it is quite unlikely that the data from a contingency table has a high degree of complexity. We expect that pairs of magazines will show interactions and when we take these into account we will not be surprised to find additional interactions for sets of three magazines. However, surely we would hope that taking account of the interactions in sets of three magazines then there might be no further interaction for four magazines. Stated differently, this means simply that we ought to be able to predict what the data for sets of four magazines would look like, granted our knowledge of the interactions by pairs and triplets. Consider the enormous saving that this would produce in the required data base. Instead of needing more than two million sets of five magazines each we could calculate any number of magazines ≥ 4 from our knowledge of all sets of three - only 1960 different sets. A straightforward statistical extension of the same general method of calculation presented in Section 38 will enable us to answer this question, which can be roughly translated into: How large must the sets of magazines be so that we can calculate the information we want for any larger number of magazines? Needless to say, it is the Boolean probability theory that permits us to make this kind of calculation.

This same problem arises in the analysis of any kind of contingency table. Suppose we have a contingency table presenting the probability of babies developing colic for five dichotomies: say parity of baby (1st or ≥2nd), diet (breast feeding or other), race of mother (Caucasian or other), education of mother (≤ 12 years of education or > 12 years of education), household income (≤ $25,000 or > $25,000). Then we can, and should, raise exactly the same questions. In fact, the causes of colic seem to be quite complex. Can we calculate the probabilities in the various logical classes by knowing two or three of these dichotomies? It is by answering such questions that we find fundamental insights into the causal factors at work in the contingency table. Section 39 will deal with this kind of question and that Section will be an appropriate one to end this book except for some final comments in Section 40.

Section 36. The Boolean Algebra of Media Data

Like all decision makers, advertisers attempt to obtain the maximum possible return for a given investment. They have major difficulties in defining the value received as a result of a specific advertising exposure and there are many fascinating problems involved with the attempt to measure values. We have no interest here in those problems. It is clear that even if the advertiser knows exactly how to measure values he or she will not be able to optimize unless there is data available concerning what can be expected in the way of exposures from specific media. This is our subject in this Section: media data. We will use magazine data as an example but data for radio shows, television shows, billboards, and so forth, are similar.

Even with this specification we will still be ignoring lots of other very interesting problems. A major example is: What should be measured for a given magazine? Possibilities are subscriptions, subscriptions plus newstand sales, readership, noting scores for advertisements, brand registration for advertisements, claim conviction for advertisements, intent to purchase - which? We will also ignore all of these questions and confine attention here to readership data but we note that the other possible measurements would be similar to readership data.

The questions are: How is magazine data presented in the industry, what are the differences among the various presentations, and how does one go from one to another presentation? It turns out that these questions can best be answered by recognizing that the problem is essentially one in Boolean algebra.

The first of the two major ways of presenting media data is called the "reach" presentation in the industry. It is equivalent to the operation ordinarily represented by a cup-shaped symbol in Boolean algebra and it is called the union. Here I will use * instead. Then, suppose we are considering two magazines, A and B. The reach of A and B together is the Boolean union of A and B, designated here as A*B, and means all those persons who saw (by the definition of readership) either A or B or both. It is equivalent to the inclusive sense of the word "or". (As, for example, when we say that a great opera star may have a wonderful voice or may be a fine actor - and we include the hoped for possibility that the star both has a wonderful voice and is a fine actor.)

The second major way in which data is presented is represented in Boolean algebra by a cap symbol and is called the intersection. For A and B, the intersection of A and B is that group of persons who read both A and B. This corresponds to the word "and". Since it is logically akin to multiplication and since I do not have a cap-symbol on my keyboard, I will simply let AB represent the intersection of A and B for our purposes.

For media data it is customary to give the data in the form of proportions. Thus, A = .197 would mean that 19.7% of the population in question read A. This means that for n magazines complete information about the readership requires (2^n - 1) pieces of information, where the -1 results because we know that the sum of all of the probabilities must add up to 1. When it is remembered that media information is needed for a variety of different sub-groups of the total population and that there are many different magazines (and huge numbers of other media) it is clear that the organizations that supply media data are only going to give it in one form. This makes the problems of translating from one form to another a very common one. So far I have only mentioned two forms and these are the two forms in which data are usually presented. We shall see shortly that there are two other forms which are likely to be of considerable interest to the advertiser.

Let us see some actual data for four magazines. Here are the reach and the intersection presentations of the data for the same four magazines:

REACH	INTERSECTION
D = .383	D = .383
E = .334	E = .334
F = .346	F = .346
G = .310	G = .310
D*E = .532	DE = .185
D*F = .556	DF = .173
D*G = .539	DG = .154
E*F = .522	EF = .158
E*G = .493	EG = .151
F*G = .520	FG = .136
D*E*F = .644	DEF = .097
D*E*G = .627	DEG = .090
D*F*G = .661	DFG = .085
E*F*G = .624	EFG = .079
D*E*F*G = .715	DEFG = .052

Thus, the last entry in the reach column means that if all four magazines are used then 71.5% of the population will read at least one of them. The last entry in the intersection column means that if all four magazines are used then 5.2% of the population will read all four of them, thus seeing four exposures of the advertisement. Notice that it is very easy immediately to recognize which of these two presentations is being used: the entry for A and B will be larger than the entries for A and B separately if it is the reach presentation and smaller than the entries for A and B separately if it is the intersection presentation.

Obviously, the two presentations are very different. Yet each of them gives the necessary 15 pieces of information so that we know everything about these four magazines. In particular, they give all of the information necessary to deduce that breakdown which is overwhelmingly of the most interest to the advertiser: the exposure distribution. The exposure distribution shows, under the assumption that a specified set of n magazines is being used, the proportion of the population which will read 0, 1, 2, . . ., n of the magazines. Note immediately that this distribution will have only (n + 1) pieces of information. This means that if one were given the exposure distribution only then one does not have complete information. In particular, one could not reconstruct either of the presentations above from the exposure distribution. However, we want to go in the reverse direction. Given either the reach or the intersection presentation, how can we deduce the exposure distribution? Here is the exposure distribution for our four magazines:

EXPOSURES	PROPORTION
0	.285
1	.304
2	.216
3	.143
4	.052
	1.000

It is evident that if one has the reach breakdown then the proportion seeing 0 magazines is 1 - D*E*F*G = .285. And if one has the intersection presentation then the proportion seeing all four magazines is DEFG = .052. But how can the whole distribution be deduced?

This is an important question in the advertising world. Suppose that the advertising decision maker cannot make some of the more refined distinctions that are implied when one speaks of the value of different classes of readers, using our four magazines as an example. Using the intersection presentation, consider DE = .185 and FG = .136. All of the readers in each of these two classes saw two advertisements. Presumably, there may be some difference in value to the advertiser of a "typical" reader in each of the two classes. There are, for example, differences in the editorial content of D and E, on the one hand, and of F and G, on the other hand. It is not difficult to imagine that such differences make a "typical" reader in the DE class more or less valuable than a "typical" reader in the FG class. But can the advertiser make a reasonable subjective estimate of this difference? Often, the decision maker does not think so. In this case, he or she might reason that each of these readers saw two advertisements and that as a first approximation it is reasonable to lump everyone who saw exactly two advertisements into one value class - surely worth more than seeing just one advertisement and worth less than seeing three advertisements. If this is so, then, the advertising executive may conclude that the exposure distribution provides the necessary basis for the best discriminations he or she can make.

Given the intersection form, it turns out that the exposure distribution can be easily found, quite neatly, by introducing some quantities designated as S_i. We define $S_0 = 1$ by definition. Then,

S_1 = the sum of all of the intersections with just one magazine

$$= .383 + .334 + .346 + .310 = 1.373$$

S_2 = the sum of all of the intersections involving exactly two magazines

$$= .185 + .173 + .154 + .158 + .151 + .136 = .957$$

and similarly for any S_i for $i \le n$ = the number of magazines in the set.

Thus, for our example $S_3 = .351$ and $S_4 = .052$.

The use of the S_i to deduce the exposure distribution is an example of Bonferroni's inequalities, which are well known in probability theory but not, I think, so well known in the advertising industry. Let P_i represent the proportion (equally well interpretable as probabilities) of the population reading i magazines. Then we can write an equation which expresses P_i as a function of the S_i. This equation requires the use of combinatorial symbols. Let ${}_nC_i$ represent the combination of n things taken i at a time, so

$${}_nC_i = n!/(i!)(n-i)! \text{ where } n! = n(n-1)(n-2)*...*(3)(2)(1).$$

Then the desired equation is

$$P_m = S_m - {}_{m+1}C_mS_{m+1} + {}_{m+2}C_mS_{m+2} - \ldots + (-1)^{n-m}{}_nC_mS_n.$$

This is a lot easier to do than it may appear. P_0 is particularly neat:

$$P_0 = S_0 - S_1 + S_2 - S_3 + \ldots + (-1)^nS_n.$$

Thus, for our example,

$P_0 = 1 - 1.373 + .957 - .351 + .052 = .285$
$P_1 = 1.373 - 2*.957 + 3*.351 - 4*.052 = .304$
$P_2 = .957 - 3*.351 + 6*.052 = .216$
$P_3 = .351 - 4*.052 = .143$
$P_4 = .052$.

I have never seen comparable equations based on the reach presentation (unions) but they are easy enough to work out. We must define $T_0 = 0$. We define T_i in exact analogy to the S_i. Thus,

T_i = the sum of the reaches of the magazines taken i at a time.

For our example, $T_1 = 1.373$ (the same as S_1), $T_2 = 3.162$, $T_3 = 2.556$, and $T_4 = .715$. Then the equation expressing P_i in terms of the T_i is

$$P_0 = 1 - T_n$$

$$P_m = -T_{n-m} + {}_{n-m+1}C_1 T_{n-m+1} - {}_{n-m+2}C_2 T_{n-m+2} + \ldots + (-1)^{m+1} {}_nC_m T_n .$$

For our example, $P_0 = 1 - T_4 = 1 - .715 = .285$, $P_1 = -2.556 + 4*.715 = .304$, $P_2 = -3.162 + 3*2.556 - 6*.715 = .216$, $P_3 = -1.373 + 2*3.162 - 3*2.556 + 4*.715 = .143$, and $P_4 = -0 + 1.373 - 3.162 + 2.556 - .715 = .052$.

If desired, one can calculate the S_i from the T_i and vice versa. To get the S_i from the T_i use
$S_m = (-1)^{m+1} T_m + (-1)^m {}_{n-m+1}C_1 T_{n-1} + (-1)^{m-1} {}_{n-m+2}C_2 T_{m-2} + \ldots {}_{n-1}C_{m-1} T_1$. m+1 m m-1

For example, from this equation $S_1 = T_1$, $S_2 = -T_2 + (n-1)T_1$, and $S_3 = T_3 - (n-2)T_2 + {}_{n-1}C_2 T_1$, and so forth. To get the T_i from the S_i we have the very pleasing result that all that needs to be done is to reverse the S's and the T's in the above equation. If you think about it, that is quite a pretty kind of symmetry.

The symmetry is preserved if you want to deduce a specific reach (union) from the intersection form or a specific intersection from the reach form. Consider obtaining a specific union from the intersection form. Suppose you want D*E*F. Treat these three magazines as if they were the whole universe and find S_1, S_2, and S_3 for these three magazines only. Here you will find $S_1 = 1.063$, $S_2 = .516$, and $S_3 = .097$. Now, the desired union is simply T_3 for these three magazines only, or $S_1 - S_2 + S_3 = .644$. Suppose you want a specific intersection, say DEF, and you are given the reach breakdown. Define the T_i over the three magazines only. Thus, $T_1 = 1.063$, $T_2 = 1.610$, and $T_3 = .644$. What you want (DEF) is obviously S_3 for these three magazines only, so DEF = $T_1 - T_2 + T_3 = .097$.

The remaining breakdown is one which is often of interest to the advertiser and is of primary interest to us since it provides the basis for applying Boolean probability theory to the data. It also contains complete information about the magazines in question. It is sufficiently important that it deserves the name "complete breakdown". It is nothing more nor less than the logical table for the set of magazines under consideration - the same logical breakdown that we have been repeatedly using throughout this book. Here it is for our sample of four magazines:

		D		D′	
		E	E′	E	E′
F	G	.052	.033	.027	.024
	G′	.045	.043	.034	.088
F′	G	.038	.031	.034	.071
	G′	.050	.091	.054	.285

The table shows the proportion of the population (usually some segment of the population) which reads each possible specific combination of magazines. This kind of breakdown is not often given by the firms which sell media information because it would take a great deal more space to present this breakdown as compared to the reach or intersection presentation. However, given this complete breakdown it is obvious that it is completely straightforward to obtain any of the other three distributions. No formulas are needed - just simple addition of the appropriate cells in the above table. For example, the proportion reading exactly three of the magazines (in the exposure distribution) is simply DEFG′ + DEF′G + DE′FG + D′EFG = .045 + .038 + .033 + .027 = .143.

Going in the other direction - from the reach or intersection presentation to the complete breakdown - is not so obvious but it is quite easy. (Remember that it is not possible to work back from the exposure distribution to the complete breakdown because it does not have complete information.) Suppose we want to obtain the complete breakdown from the intersection presentation. The last entry in the intersection presentation is the proportion of the population who have read all the magazines in the set: DEFG = .052. This is the entry in the upper left corner of the logical table. Then we get four more entries from the three-magazine intersections thus: DEF - DEFG = .045 = DEFG′ and EFG - DEFG = .027 = D′EFG. Now we have the top three entries in the column showing DE but DE = .185 and subtracting the sum of the three from this gives DEF′G′ = .050. We have three entries for EF so subtracting the sum of these three from EF = .158 gives D′EFG′ = .034. If you follow this process out you will find that you can obtain six more entries from the intersections of pairs of magazines and then four more from the individual magazines and, finally, the entry for D′E′F′G′ from the requirement that the sum of all of the entries must equal one. A similar procedure starting with the reach presentation begins with 1 - D*E*F*G = .285 as the lower right-hand entry in the complete breakdown and builds up to complete the logical table by using, successively, the reaches by three magazines, then by two magazines, and then by single magazines. The final entry, D*E*F*G = .052 is obtained from the requirement that the sum of all of the entries must equal one.

The point of the preceding paragraph is certainly not that we want generally to find the logical table (the complete breakdown) by somewhat laborious hand calculations. It is intended merely to show that the logical process is quite straightforward. Naturally, in our analysis we will want to obtain any needed numbers by computerized procedures. The most direct approach to this end is to write equations in terms of what we may call here the Boolean variables of the set of magazines. What we want is one variable for each of the cells of the table except that we do not need one for the cell representing the proportion which did not read any of the magazines. In order to have a convenient mnemonic system for writing these variables I use small letters with the convention that only those letters actually written were magazines actually read. Thus, let dg be the proportion of the population which reads both D and G and did not read E or F; let e be the proportion of the population which read only E; defg is the proportion which read each of the magazines; and so forth. These variables are quite useful and we will use them when we illustrate the use of linear programming for some media problems with partial information.

Now return to the reach and intersection presentations and note that we can write an equation, using our variables, for each piece of information given. Thus, for the reach presentation we have that E*F = .522. In terms of our proposed variables this includes every one which has an e or an f in it (or both). So:

$$e + f + de + df + ef + eg + fg + def + deg + dfg + efg + defg = .522 .$$

Note that we have fifteen pieces of information. Hence, we have fifteen equations and since we have fifteen unknowns we can solve them to find the complete breakdown. Similarly, from the intersection form we have EF = .158. In terms of our variables this includes every variable which has both e and f in it:

$$ef + def + efg + defg = .158 .$$

Again we have 15 equations in 15 unknowns so we can solve to determine the complete breakdown.

To talk of fifteen equations in fifteen unknowns sounds extremely threatening for hand calculations but we will ordinarily be using a computer. In any case, the equations are of such a structured form that they can be solved quite easily - even by hand. Thus, for example:

$$d = D*E*F*G - E*F*G \qquad e = D*E*F*G - D*F*G$$

$$f = D*E*F*G - D*E*G \qquad g = D*E*F*G - D*E*F$$

These equations are based on the general fact that if you have a union of j magazines and if you subtract from it a union of i magazines which is a subset of the j magazines then you are left with all possible combinations of the (j - i) magazines. Thus, we can continue. For example, D*E*F*G - D*G = e + f + ef. Since we already have e and f we can now solve for ef. Or we can work it out symbolically, once and for all:

$$ef = -D*E*F*G + D*F*G + D*E*G - D*G$$

and

$$de = -D*E*F*G + D*F*G + E*F*G - F*G$$

and similarly for the other four with just two letters. In order to find the terms with three letters we proceed as follows. Consider efg. We know that

$$D*E*F*G - D = e + f + g + ef + eg + fg + efg .$$

Since we know all of the terms here except for efg it follows that we can solve for efg. Doing it once and for all:

$$efg = D*E*F*G - D*E*F - D*E*G - D*F*G + D*E + D*F + D*G - D .$$

The other variables with three letters have the same pattern so they can be found in the same way. We now have 14 of the unknowns and the last one, defg, can be found by subtracting the sum of the 14 from D*E*F*G or by $T_1 - T_2 + T_3 - T_4$.

It is easier to solve for our Boolean variables from the intersection breakdown. Consider solving for d. Let us confine our attention to the subpopulation who read D. If we made the exposure distribution for this subpopulation then P_1 would obviously be those who had read only D, which is d. Therefore, we can use the formulas given previously to find P_1, but defining our S_i in terms of the subpopulation. It is perhaps most perspicuous written this way:

d = D - Σ(2's with D) + Σ(3's with D) - Σ(4's with D)

= .383 - .512 + .272 - .052 = .091.

Similarly,

de = DE - Σ(3's with DE) + Σ(4's with DE)

= .185 - .187 + .052 = .050 .

This is a sufficient introduction to the Boolean algebra of media data for our immediate purposes. We will now briefly present some of the non-Boolean procedures that are often used to help the advertising decision maker.

Section 37. Some Useful non-Boolean Approaches to Media Data.

The basic problem that almost always obtrudes on decision making in advertising is that the decision makers hardly ever have anything approaching complete information about the choices available to them. It is worth some space to make sure that the fundamental difficulty is clear. In courses in basic statistics the students are introduced to joint probability tables. When two different events are independent such tables are particularly easy to determine. Consider the simultaneous rolling of a pair of dice with the event in question being the occurrence of a 7 or not and the drawing of a card from a well-shuffled deck with the event being the selection of a heart or not. The four possibilities, with the associated probabilities can be conveniently presented thus:

	DICE		
	7	not 7	
CARDS: HEART	1/24	5/24	1/4
NOT HEART	3/24	15/24	3/4
	1/6	5/6	

Here, since the two events are clearly independent, the joint probabilities of the two events are simply the products of the probabilities of the individual events.

In contrast, consider the data for readership of two magazines. The data can be put in exactly the same form but the events are never independent. Here are some old data for Good Housekeeping and Readers Digest, measured as adult readership:

	GH	not GH	
RD	.059	.244	.303
not RD	.055	.642	.697
	.114	.886	

The entries are proportions - equally interpretable as probabilities - and the two magazines are definitely not independent events. Thus, if they were independent then the proportion reading both magazines would be .303*.114 = .0345 whereas it is actually .059. Remember that these proportions are applied to large numbers. The total number of adults is about 115,340,000 for the specific class covered by these data. Therefore, if the advertiser had already decided to use Readers Digest then he or she would anticipate that Good Housekeeping would add about 9,150,000 adults to the 34,948,000 reached by Readers Digest. Because of the lack of independence it only adds 6,340,000. This is a very considerable difference and would very much increase the cost per adult reached.

This granted, the most important fact thus illustrated is that because of the lack of independence it is impossible to calculate the .059 entry in the table. Therefore, it had to be measured. To complete the table for the independent events in our example we needed only two pieces of information: 1/6 (probability of 7) and 1/4 (probability of a heart). But for the magazines we need three pieces of information: .114 (the probability that an adult reads Good Housekeeping), .303 (the probability that an adult reads Readers Digest, and .059 (the probability that an adult reads both of the magazines). The difference may not seem very great but this is because we are dealing with just two events. If we consider n different events then to complete the corresponding table showing the probabilities of all of the possible combinations we need:

if independent : n pieces of information

if not independent : $2^n - 1$ pieces of information.

As n gets large this difference becomes enormous. For 20 magazines we need more than 1,000,000 pieces of information and for 30 magazines we need more than 1,000,000,000. These would be required in order to have complete information about all of the possible combinations. Since it is obviously impossible to obtain this amount of information it follows that we are almost always in the position of having only partial information concerning media. Large advertisers often use 30 or more magazines over the course of the year and they also will use a variety of other media: AM radio, FM radio, TV, billboards, and so forth. Complete information for such advertisers would require an enormous logical table, which is, in fact, totally impossible to obtain. Still, these advertisers are anxious to make "rational" decisions. How can they hope to do so when they only know a relatively tiny fragment of the total information about the effects of their various media selections?

There are various methods which can be used to enable the advertising decision makers to base their decisions on reasonable approximations to the total information, which is unavailable. The most appropriate one of the possible alternatives usually depends heavily on the details of the advertising situation and an entire book would be required to present all of the different kinds of subjective valuations that can be involved in advertising decisions. This is certainly not an appropriate goal for our presentation. Instead, we will illustrate a few of the methods in terms of specific objectives. Our intention is to emphasize that Boolean probability theory applied to these media problems, and discussed in the next Section, is a much more useful approach than the methods we will consider in this Section.

A. Finding bounds on probabilities; primarily linear programming.

The general idea underlying the various methods is to use the available information in some reasonable way that will lead to a prediction of the remainder of the information, which is otherwise not known. For example, suppose an advertiser is in the relatively common situation of knowing the reaches of individual magazines and the duplications of pairs of magazines (either as reaches or as intersections) but knows nothing else. Suppose the advertiser is considering using four magazines: A, B, C, and D. The following information is available (intersections):

A	:	.161	AB	:	.075
B	:	.229	AC	:	.049
C	:	.220	AD	:	.090
D	:	.383	BC	:	.081
			BD	:	.122
			CD	:	.096

These are actual, but old, data. Suppose the advertiser's objective is simply reach. This means that the advertiser does not discriminate among the households according to the number of exposures they have received. Rather, he or she ascribes a value of one to each household reached. The argument underlying this is the ancient industry adage: "If you don't reach them, you can't influence them."

How much reach might be achieved with the four magazines? We are lacking five pieces of information, as compared to having total information. We can write a linear program in terms of the small-letter variables we introduced earlier. With these variables, "a" means the household only saw A, "dg" means the household saw exactly, and only, D and G, and "agh" means the household saw A, G, and H. We can write a linear program conveniently by putting the variables at the top and the coefficients for each equation beneath:

a	b	c	d	ab	ac	ad	bc	bd	cd	abc	abd	acd	bcd	abcd		
1				1	1	1				1	1	1		1	=	.161
	1			1			1	1		1	1		1	1	=	.229
		1			1		1		1	1		1	1	1	=	.220
			1			1		1	1		1	1	1	1	=	.383
				1						1	1			1	=	.075
					1					1		1		1	=	.049
						1					1	1		1	=	.090
							1			1			1	1	=	.081
								1			1		1	1	=	.122
									1			1	1	1	=	.096

Objective function:

1	1	1	1	1	1	1	1	1	1	1	1	1	1	1	=	z

We have ten pieces of information and there is one equation for each. The objective function, commonly denoted "z", is the sum of all of the 15 small-letter variables. Using linear programming, we can either maximize or minimize "z". In other words, what is logically the maximum possible reach that could be achieved with these four magazines? Similarly, what is logically the minimum possible reach that might be achieved? We want to find both of these solutions. This is such a small linear program that it can be solved almost instantaneously. Here are the two solutions:

MAXIMUM	MINIMUM
a = .071	d = .126
b = .107	ab = .026
c = .124	ac = .043
d = .231	ad = .039
ad = .015	bc = .079
bd = .015	bd = .075
cd = .015	cd = .092
abd = .026	abc = .002
bcd = .032	abd = .047
abcd = .049	acd = .004
Reach = .685	Reach = .533

Notice that all of the known information is respected in the two solutions. Thus, in the maximum solution A would be the total of all of the small-letter variables that include "a", here .071 + .015 + .026 + .049 = .161, as it should. In the minimum solution A will be .026 + .043 + .039 + .002 + .047 + .004 = .161.

What we have now discovered is that using these four magazines guarantees at least a reach of .533 and we also know that it would be impossible to achieve a reach greater than .685. But ordinarily the advertising decision maker will want a single "best" estimate of the reach that will be achieved with these four magazines - so that he or she can compare the results with other selections of magazines and with the various costs of purchasing space in the selected magazines. How can we get a single "best" estimate? For this purpose we have a result due to George Polya (American Mathematical Monthly; 1950). Suppose we know only that a variable of interest, "x", falls within a range: $a \leq x \leq b$, where $0 < a < b$. If we want to estimate the value of x, say by an estimate $\tilde{x}$, then Polya points out that we might want to minimize the extreme error, $\tilde{x} - x$, or the relative error, $(\tilde{x} - x)/x$. The former choice leads to

$\tilde{x} = (a + b)/2$ (the arithmetic mean of the two limits) and the latter leads to $\tilde{x} = 2ab/(a + b)$ (the harmonic mean of the two limits). The latter is almost always preferable. So, to get a single "best" estimate we will follow Polya's argument and estimate the reach that will be achieved as the harmonic mean of the two limits we have discovered: reach = (2*.685*.533)/(.685 + .533) = .600. For these four magazines I have the measured data showing the reach of all four magazines actually to be .607 so our estimate is, as very often, quite close to the reality - which would usually be unknown or else we would not be using linear programming.

This procedure, which I call Boolean programming is logically impeccable but it suffers from one serious difficulty. This is that the variables are all of the possible combinations of the magazines and this means that if we are working with, say, 20 magazines we would need more than 1,000,000 variables and it is totally impracticable to handle LP's routinely with this many variables. Even in this day of lightning-fast calculations and super-computers this kind of linear programming would exceed the practical limits of computation. And remember that ordinarily the decision makers would want such analyses on many different sets of 20 or more magazines. What is to be done?

The answer to the preceding question is that we can use the exposure distribution to construct a different linear programming formulation which has a very modest number of variables. We need again to refer to some of the quantities we introduced in the Section on media data. Specifically, S_1 was defined as the sum of the individual magazine reaches and S_2 is defined as the sum of the intersections of the magazines two at a time. Thus, for our preceding example we gave the individual magazine reaches and the intersections by two magazines at a time. For these data we find $S_1 = .993$ and $S_2 = .513$. The exposure distribution for these four magazines is

Exposures	Probability
0	.393
1	.336
2	.168
3	.091
4	.012

Note that this exposure distribution cannot be calculated from the data given for these four magazines. I used additional data, not quoted here, to calculate this distribution. If we calculate the mean of this distribution we find that the mean number of exposures is $.993 = S_1$. If we calculate the variance we find that it is 1.032951. Alternatively, we notice that variance + mean^2 = 2.019 = $S_1 + 2S_2$. These two relationships always hold true:

For the exposure distribution: mean = S_1

variance + mean^2 = $S_1 + 2S_2$.

We can summarize these relationships by stating that knowing the reaches of the individual magazines and the intersections of the magazines by pairs - and this is the information most often available - signifies that we know the mean and the variance of the exposure distribution.

How can we use linear programming on an exposure distribution? Consider our example. Given only the reaches and intersections by pairs we cannot calculate the actual exposure distribution. Without knowledge of the intersections by threes we do not know how many households got three exposures and we also do not know how many households got four exposures. Therefore, we also don't know how many households got zero exposures. But what we **do** know can be written like this:

Exposures	Proportion
0	p_0
1	p_1
2	p_2
3	p_3
4	p_4
	1

Concerning this distribution we know that:

$$\Sigma p_i = 1$$
$$\text{and } \Sigma i p_i = S_1 \text{ or } p_1 + 2p_2 + 3p_3 + 4p_4 = .993$$
$$\text{and } \Sigma i^2 p_i = S_1 + 2S_2 \text{ or } p_1 + 4p_2 + 9p_3 + 16p_4 = 2.019.$$

These are the "Subject to" conditions. Subject to these we want to maximize and/or minimize the reach, which is simply 1 - P_0. This is obviously a linear programming problem.

For the specific case where the objective is reach I have worked out an analytical solution to the linear programming problem. In other words, the maximum and minimum reaches can be found by straightforward calculations without having to do the linear programming analysis. Here is the procedure, where n is the number of magazines:

MAXIMUM : The maximum reach is either $S_1 - (2S_2/n)$ or 1, whichever is smaller.

MINIMUM : Let k be the smallest integer > $1 + (2S_2/S_1)$. Then the minimum reach is

$(2/[k(k-1)])([k-1]S_1 - S_2)$.

For our example the maximum reach is .993 - 1.026/4 = .7365. For the minimum we calculate that $1 + (2S_2/S_1) = 2.033$ so k = 3 and the minimum reach is (2/3*2))(1.986 - .513) = .491. Once again we take the harmonic mean to obtain an estimate of the actual reach which will be achieved. Here the harmonic mean = .5892, in error by 2.9% as compared to the actual reach of .607. This is a somewhat larger error than we usually obtain.

This method is practicable for any number of magazines. The number of variables in the linear programming formulation will be n + 1, where n is the number of magazines. Further kinds of partial information can be incorporated if such is available. Any objective function based on the exposure distribution can be used - and almost all objective functions in practice are of this kind. Only in the case where the objective function is reach can the algebraic solution given above be used. In all other cases it would be necessary to solve the LP problem. Of course, it is quite an easy one to solve since there are usually only a few restrictions.

One more point should be made. Generally, this approach will not give as sharp limits as the previous approach. This is because the exposure distribution does not contain complete information. Thus, in our example, the information available gave ten restricting equations whereas the exposure distribution had only three. This obviously represents less precise statements of the information and this is reflected in less sharp upper and lower limits. However, this is ordinarily partially compensated for by the process of taking the harmonic mean.

It is worth calling attention to a characteristic of these linear programming solutions that leads decision makers to try other approaches. Here are the specific LP solutions for our example:

Exposures	Maximum	Minimum
0	.2635	.509
1	.651	
2		.480
3		.011
4	.0855	
	1.0000	1.000

The characteristic in question is evident in both distributions but is somewhat more striking for the maximum. This is that there is simply no resemblance between the maximum distribution found by LP and any ordinary exposure distribution found in the real world. It is almost unimaginable that we would find an actual exposure distribution with a large probability (.2635) of no exposures and a very large probability (.651) of exactly one exposure, zero probability of two or three exposures and a fairly large probability (.0855) of four exposures. Actual distributions do not arise with such erratic entries. The same holds true of the minimum distribution. Essentially, this says that half of the households get zero exposures and the other half get two exposures - with none getting one exposure. Reality rarely shows this kind of lumpiness. The exposure distribution for our four-magazine example is given above and this one is fairly typical. It is unimodal - with the mode at zero exposures - and the probabilities go down relatively smoothly over 1, 2, 3, and 4. The LP solutions almost never demonstrate this kind of behavior.

B. Jaynes' Principle.

The conclusion of this line of argument can be stated quite simply: one would like to know the maximum and minimum reaches achievable but while confining the possibilities to "reasonable" exposure distributions. This objective can be stated quite simply but doing it is another thing entirely. The problem is that it is quite difficult to define mathematically what we mean by a "reasonable" exposure distribution. This granted, there are quite a number of different such definitions that are sometimes used to provide a basis for decision making in advertising. In this Section we will illustrate one fairly standard such procedure. Let us first note that there are two major approaches here. One is to define "reasonable" and then attempt to find the maximum and minimum from among the "reasonable" possibilities. For example, perhaps we might try defining "reasonable" to mean a unimodal distribution with the measurements smoothly increasing until the mode and then smoothly decreasing after the mode. Perhaps if we could do this cleverly enough we would be able to use one or another known kind of programming to find the maximum and minimum within this defined group of possibilities. The second approach attempts, instead, to define what we mean by "rational" behavior. If this can be defined mathematically then we may be able to directly find the "rational" conclusion to be drawn from given partial information. In this case there will be no need for finding a maximum and a minimum and then from them constructing a "best" estimate. The "best" estimate will be found immediately by the mathematical argument. Our example is of this second type.

The approach that we are going to use as our example is a very basic approach proposed by a well-known physicist, E. T. Jaynes, for dealing with any decision problem with only partial information. So doing is called, for obvious reasons, following Jaynes' Principle. First, let us simply state Jaynes' Principle and then try to understand the reasons that underlie the quite mysterious recommendation:

> If partial information is available concerning a probability distribution then determine that probability distribution which meets the requirements of the partial information and which maximizes
>
> $$-\Sigma p_i \log p_i$$
>
> and assume that this is the unknown distribution and act accordingly.

The logs can be to any base but often they are taken to the base 2 and in this case the quantity being maximized is considered to be measured in "bits". The quantity in question - that is being maximized - is otherwise known as the entropy of the distribution.

It is certainly not obvious why Jaynes recommended that one should act in this way. However, his argument has considerable appeal and books have been devoted to it. One of the most common kinds of decision problems, arising in almost every decision making area, is how to deal with situations where there is only partial information about the key probability distributions involved. There have been very widespread attempts to utilize Jaynes' Principle in many of these areas. For example, there are articles in the financial literature considering the merits of using Jaynes' Principle in portfolio management. We would like to get some idea of the underlying rationale before we illustrate the applications of the Principle to decision making in advertising.

In order to gain some understanding it is convenient to consider the parlor game called "Twenty Questions". As it is often played, one member of a group (the Chooser) will privately pick and keep in mind some person - real or fictional, living or dead - and the other members of the group have up to 20 questions, each having a "Yes" or "No" answer, which they can ask in order to identify the selected person. If the group cannot identify the person in 20 questions then the Chooser wins, otherwise he or she loses. The reason that the game of "Twenty Questions" is relevant to Jaynes' Principle is that there is a direct relationship between the number of questions that "ought" to be required and the entropy.

In order to illustrate this let us take an extremely simple version of the game. Suppose that I am going to secretly select one of the letters

A B C D E F G H

and that you must guess which letter I have selected by asking me a series of yes-and-no questions. I think that if you consider this simple problem you will quickly discover that you can always determine which letter I have selected with exactly three questions. Thus, it might go

Is it A, B, C, or D?	No
Is it E or F?	Yes
Is it E?	No

Therefore, it is F. In doing this you used no assumptions about the likelihood with which I might select one or another letter. In effect, you assumed that I was equally likely to select any letter.

Suppose that you had information suggesting that I had a predilection for selecting one letter. Specifically, suppose that I am now selecting from among nine letters instead of the eight above:

A B C D E F G H I

but that you know that half the time I select A, with the remaining letters being equally likely. Then commonsense suggests that the first question should be: Is it A? Half the time I will say "Yes" and you will have succeeded with one question. The other half of the time I will say "No" and then you are back to the first case we considered, which we already know requires three additional questions. Therefore, the expected number of questions will be .5(1) + .5(4) = 2.5. You see that less questions are required, even though there are more letters, because of taking into account my preferences.

What has this got to do with Jaynes' Principle? Let us calculate the entropy of the two situations. For the first one we had 8 possible letters, each with probability 1/8. We will use logs to the base 2 and

$$-(1/8)\log(1/8) - \ldots \text{(eight times)} = -\log(1/8) = 3$$

For the second case:

$$-(1/2)\log(1/2) - (1/16)\log(1/16) - \ldots \text{(eight times)} = -(1/2)\log(1/2) - (1/2)\log(1/16)$$

$$= .5 + 2 = 2.5\,.$$

Thus, we see that the entropy, measured in bits, is the number of Yes-or-No questions required to find out what happened for a given probability distribution.

The more questions that are required to find out what happened, the greater is one's uncertainty concerning the given distribution. So, what Jaynes is recommending is that we should always assume that the probability distribution is that one which maximizes our uncertainty, granted the partial information which is available. Put differently, he is saying that we should always assume that that probability distribution is true which makes the usefulness of our partial information a minimum. This seems to be a very good way to take a conservative stance in decision problems with partial information.

Let us continue our example of four magazines and let us assume that all we know is that there are from 0 through 4 exposures, that the sum of the probabilities is one, that S_1 = .993, and that S_2 = .513. We can find the maximum entropy distribution with these characteristics but even though the mathematics involved is not overly difficult it is certainly true that it is generally considered a part of relatively advanced applied mathematics. We need to use Lagrangian multipliers and related methods. Therefore, we will not illustrate here how the desired maximum entropy distribution is obtained. However, it is a good example of the way the application of Jaynes' Principle always leads to "reasonable" distributions. (See Appendix B for the procedure to follow.) The maximum entropy distribution for the four magazines is:

Exposures	Probability
0	.39285
1	.32994
2	.18708
3	.07162
4	.01851
	1.00000

In fact, in this case the maximum entropy distribution is quite a bit like the actual distribution but this is not true in general.

We will illustrate the other kind of application of Jaynes' Principle to partial information in advertising decision problems. Specifically, suppose that for magazines D, E, and F we know

$$D = .383 \qquad DE = .185$$
$$E = .334 \qquad DF = .173$$
$$F = .346 \qquad EF = .158\ .$$

We are lacking one piece of information, DEF. We can immediately construct the logical table thus:

	D		D′	
	E	E′	E	E′
F	x	.173 - x	.158 - x	.015 + x
F′	.185 - x	.025 + x	x - .009	.453 - x

Notice that we immediately have bounds on x as $.009 < x < .158$ to ensure that none of the entries in the table are negative. To use Jaynes' Principle in estimating x, we want to maximize

$$-\Sigma p_i \log(p_i) = -[x\ln x + (.173-x)\ln(.173-x) + \ldots + (.453-x)\ln(.453-x)\ .$$

Taking the derivative with respect to x and simplifying gives

$$x(.025 + x)(x - .009)(.015 + x) = (.185 - x)(.173 - x)(.158 - x)(.453 - x)\ .$$

Note that the quickest method for obtaining the logical table given above will be discussed in the next Section and also note that we will see this identical equation again shortly. The equation can be solved directly in this form by computer, using numerical methods if convenient. I find the root x = .09996483. This gives the logical table

	D		D′	
	E	E′	E	E′
F	.100	.073	.058	.115
F′	.085	.125	.091	.353

Since the actual value of def is .097, the reconstructed logical table is very close to the actual values. Furthermore, the agreement of the exposure distributions is very close:

Exposures	Actual	Estimated
0	.356	.353
1	.322	.331
2	.225	.216
3	.097	.100

However, it should be emphasized that from our standard logical table, written without knowing DEF, we could write the resulting exposure distribution and apply Jaynes' Principle directly to that. The present point is that the resulting value of x will not be identical with the x = .0999648 found from applying Jaynes' Principle to the logical table. It will generally be close in value but it is a different mathematical problem and has a different answer - here, x = .0997646.

This is a good example showing why decision makers in advertising prefer estimation methods which give "reasonable" distributions. The general effect is to enhance confidence in the resulting decisions. This may not be a valid argument but it certainly is a congenial one.

C. Changing N to make magazines become independent random variables.

Consider how the data on magazine readership (or any of the other media) is obtained. Let us take as an example data which is most solidly supported, for example circulation data for magazines provided by Daniel Starch.[1] On the one hand, the reported circulation figures for major consumer magazines can be taken to be very reliable. But, on the other hand, the size of the population from which the households subscribing, or buying the magazine at a newstand, are a sample is subject to all kinds of uncertainty and even an exact definition of the characteristics of this unknown population is almost impossible to achieve. Therefore, it is a reasonable possibility that the number of households in this undefined population is quite likely to be different than the N that Starch used in calculating the reaches of the various magazines.

Take the data for Readers Digest (R) and Life (L). It is beyond challenge that the readership of R was, as given by Starch, 11.833 million, the readership of L was 5.792 million, and the readership of both (RL) was 2.415 million. But what we are suggesting here is that the fourth entry in the standard 2x2 logical table is highly dubious:

	R	R′
L	2.415	3.377
L′	9.418	x

Suppose, therefore, that we calculate x to be the number of households in the population that did not read either R or L such that R and L were independent random variables, meaning that

$$2.415 = \frac{5.792 * 11.833}{15.210 + x}$$

which gives x = 13.175. In other words, the total population relative to these two magazines, construed as independent, is 28.385 million. Using this as N for the total population we find that R = 11.833/28.385 = .416875, L = 5.792/28.385 = .204051, and RL = 2.415/28.385 = .08508 and, finally, .08508 = R*L within slight rounding errors. In other words, with this value of N selected then R and L are acting as random and independent variables.

Is this value of N "correct" in any sense of the word? Perhaps, but we certainly have no argument to offer in favor of it - at least, yet. At the time when these data were collected, both R and L were considered, and called, "mass" magazines and it was often argued that the population of relevant households was likely to be characterized by the fact that these households liked "mass" magazines and would often subscribe to several of them. We will be satisfied to add one more "mass" magazine to our table, Saturday Evening Post (S). Starch data showed that the reach of S was 5.551 million and that

SL = 1.123 million and SR = 2.399 million. If we reconstruct the table for S and L as we did above for R and L, we find:

	L	L′
S	1.123	4.428
S′	4.669	18.410

Here S = 5.551/28.630 = .1938875, L = 5.792/28.630 = .202305, SL = 1.123/28.630 = .039225 and SL = .039225 = S*L. Reconstructing the table for S and R, we find

	R	R′
S	2.399	3.152
S′	9.434	12.395

with S = 5.551/27.380 = .202739, R = 11.833/27.380 = .432177, SR = 2.399/27.380 = .087619 and SR = .087619 = S*R.

One cannot avoid noticing the similarity between the three values of N that we obtained. This strongly suggests that perhaps one value of N is adequate for at least these three "mass" magazines and perhaps for most of the so-called "mass" magazines. Suppose we want one value of N for all three of these magazines. The most straightforward way to accomplish this is to minimize the sum of the squares of the errors that will result for the duplications in the three tables. Thus, minimize

$$\left[\frac{5.792*11.833}{N} - 2.415\right]^2 + \left[\frac{5.551*5.792}{N} - 1.123\right]^2 + \left[\frac{5.551*11.833}{N} - 2.399\right]^2$$

The minimizing value is N = 28.023. Using this value of N we obtain the following logical table:

	R		R′	
	S	S′	S	S′
L	.485	1.930	.638	2.739
L′	1.914	7.504	2.514	10.299

The reaches are: L = .206687, R = .422260, S = .198087; the duplications are: RL = .086179, RS = .085608; LS = .040074; and the triplication is RLS = .017307. The effects of our change in N are that the magazines do not quite function like independent variables but they are very close: thus, for example, RLS = .017288, RL = .087276, and RS = .083644. We did not make any sophisticated effort at allocating the changes in N so that they would have minimal effect on the values. But without making such an effort there is fairly strong evidence that these magazines may be approximately functioning as independent variables if we use a more realistic value of N.

This kind of possibility is very much suggested by outstanding success in finding a value of N for women's magazines which made six such magazines act like independent variables and similarly to find a unique value of N for five men's magazines that converted the data into that of independent random variables. It is obviously a very great saving in calculations and a vast improvement in insight if this kind of change can be accomplished.

D. An interesting and potentially useful way to utilize regression analysis on partial information.

Multiple regression analysis can be used directly on the partial information of a media decision problem with the otherwise unexpected result that the analysis will produce estimates of the unknown information. Thus, suppose, as is often the case, that the available partial information consists of the reaches and the duplications of the given magazines. Then by using multiple regression analysis we can construct complete logical tables that will give estimates of each of the other unknown quantities, including the proportion of persons reading any three of the magazines, any four of the magazines, and so forth. The analytical procedure sometimes seems elusive but it is, in fact, a completely straightforward application of the standard methods of multiple regression analysis.

In order to understand the procedure to be followed, consider the form that the original information - obtained by one of the sellers of such information - would take, for magazines A, B, C, and D, and assuming that the data is what is called household readership:

Household	A	B	C	D
1	0	1	0	0
2	1	1	0	1
3	1	0	1	1
. . .				
2000	0	1	0	1

Here 1 means the household reads the magazine and 0 means the household does not read the magazine. We would then count the number of 1s for each magazine. Suppose there are 322 for magazine A. Then dividing by the total number of households, 2000, we would find that the reach of A was rA = .161. Similarly, we might find that r_B = .229, r_C = .220, and r_D = .383. (These are the values used as an example of Boolean programming, above.) Notice that since all of the entries are either 0 or 1 it follows that $\Sigma x = \Sigma x^2$ for each magazine, where x stands for a measurement in the magazine's column. For purposes of multiple regression analysis we also need the sum of the products of the two entries in the same row (the same household) for each pair of magazines. For example, suppose the sum of the products of the measurements in the same row for A and B is 150. Then dividing this total by 2000 gives the duplication of A and B as d_{AB} = .075. Assuming our data is that from which we obtain the summary statistics given for magazines A, B, C, and D in our example of Boolean programming, we find d_{AB} = .075, d_{AC} = .049, d_{AD} = .090, d_{BC} = .081, d_{BD} = .122, and d_{CD} = .096. (The reason for the division of all of these quantities by the total number of households, 2000, will become clear shortly.)

With the information we are given we can construct the logical table for any two of the magazines. Suppose we have decided to use magazines A and B. Then we can calculate the proportion of the households that will fall into each of the four possible logical classes that result:

	A	A′	
B	.075	.154	.229
B′	.086	.685	.771
	.161	.839	

Here we know the reach of B is .229 and the duplication of A and B is .075. This gives us the first row of the logical table. Then from the reach of A being .161 and the duplication of A and B being .075 we can get the first column of the table. The entry for A'B' is obtained by subtraction, knowing that the sum of the four entries in the table must equal 1. This table shows the proportion of households that will fall into each of the four logically possible classes if A and B are jointly used.

Now we begin the multiple regression analysis. Suppose we want to consider adding magazine C to A and B. This means that we want to find a linear equation that relates a given household's response to C to that household's responses to A and B. In short, we want a linear predicting equation of the form: C = a + bA + cB, where the small letters a, b, and c are simply the usual designations of the parameters of this equation. In other words, there is no relationship between a, b, and c on the one hand and the fact that the magazines are A, B, and C on the other hand - beyond the fact that a, b, and c are the parameters in the stated predicting equation relating household behavior towards C as a linear function of that household's behavior toward A and B. The way we determine equations to find a, b, and c is by minimizing the sum of the squares of the errors we will make by using our predicting equation. We want to minimize $\Sigma(C - a - bA - cB)^2$ with regard to a, b, and c. Taking the partial derivatives of a, b, and c and slightly rearranging the results gives the standard three normal equations for our proposed equation:

$$\begin{aligned} na + b\Sigma A + c\Sigma B &= \Sigma C \\ a\Sigma A + b\Sigma A + c\Sigma AB &= \Sigma AC \\ a\Sigma B + b\Sigma AB + c\Sigma B &= \Sigma BC \end{aligned}$$

These equations already incorporate the simplification that for these data, for example, $\Sigma A = \Sigma A^2$. n is the total number of households and it is obvious that we can simplify the equations by dividing everything through by n. This convenient division is what leads to the same equations being written in terms of the usual reaches and duplications for media:

$$\begin{aligned} a + br_A + cr_B &= r_C \\ ar_A + br_A + cd_{AB} &= d_{AC} \\ ar_B + bd_{AB} + cr_B &= d_{BC} \end{aligned}$$

Therefore, the specific equations that we want to solve are, taken directly from the summary data:

$$\begin{aligned} a + .161b + .229c &= .220 \\ .161a + .161b + .075c &= .049 \\ .229a + .075b + .229c &= .081\ . \end{aligned}$$

Using Mathematica, the solution of these three equations, to eight places, is:

a = .17415867
b = .05492636
c = .16156414 .

Our predicting equation, therefore, is C = .17415847 + .05492636A + .16156414B. If A = 0 and B = 0 then C = a = .17415847; if A = 0 and B = 1 then C = a + c = .33570007; if A = 1 and B = 0 then C = a + b = .22908503; and if A = 1 and B = 1 then C = a + b + c = .39064917. Therefore, for each of the four logically possible classes resulting from A and B we now know the probability that a household of the given class also reads C. We have:

A	B	From A,B	C	With C	With C′
0	0	.685	.17415867	.1193	.5657
0	1	.154	.33570007	.0517	.1023
1	0	.086	.22908503	.0197	.0663
1	1	.075	.39064917	.0293	.0457

Thus, the calculation for the first row is that .685(.17415967) = .11929869 proportion of the households that read neither A nor B did read C and the remainder, .685 - .11929869 = .56570131 also did not read C. Therefore, from the last two columns in the table we have the logical breakdown for A, B, and C:

	A		A′	
	B	B′	B	B′
C	.0293	.0197	.0517	.1193
C′	.0457	.0663	.1023	.5657

We have often had occasion to verify that for Boolean tables it was true that adfg = ebch in our usual notation. Here we find that afgd = .00002371 and ebch = .00002633. They are not equal and there is no reason why they should be. The method of construction of the logical table (multiple regression analysis) does not in any way imply that these two quantities should be equal.

Notice that so far we have not done anything that we could not have done with Boolean analysis. Using the Boolean approach and working directly from the input data for A, B, and C we can find the following entries in the logical table, in terms of "a" = ABC:

	A		A′	
	B	B′	B	B′
C	a	.049 - a	.081 - a	.090 + a
C′	.075 - a	.037 + a	.073 + a	.595 - a

Here we are entitled to find "a" such that afgd = ebch or, writing the products as an equation = 0 we have

$$a(.037 + a)(.073 + a)(.090 + a) - (.075 - a)(.049 - a)(.081 - a)(.595 - a) = 0 .$$

Using Mathematica, I find that a = .029948552 is the desired root of this equation. Using this value of "a", we have the completed logical table:

	A		A′	
	B	B′	B	B′
C	.0299	.0191	.0511	.1199
C′	.0451	.0669	.1029	.5650

with afgd = .000024759 and ebch = .000024759. There is not much difference between the two logical breakdowns even though they are based on completely different arguments. With regard to basing our estimate of the one unknown entry, "a", on the equation afgd = ebch we may note the following. In order to pursue the Boolean analysis of this example we need to introduce the variables u = AB, v = AC, and w = BC. Having done so we can find eight equations like:

$$\frac{uvwABC}{a} = \frac{u'v'w'A'BC}{.081 - a} = \ldots = \frac{u'v'w'A'B'C}{.090 + a} = V$$

Four of these have a term involving +a in the denominators and four have negative values of "a" in the denominators. The product of the four with positive "a" values has the same numerator as the product of the four with negative "a" values. Equating the two products denominators gives the equation afgd = ebch. Therefore, finding the value of "a" as we did here is a completely orthodox Boolean argument.

Let us continue our multiple regression analysis after this brief return to Boolean analysis. The three magazines A, B, and C produce eight logical classes and our next step will be to deduce a predicting equation for D that will be based on these eight classes. The predicting equation will be of the form D = a + bA + cB + dC. Needless to say the values of the four parameters are not the same as any previous value. It is convenient to use the same letters to represent the parameters because it makes it easy to keep track of the logic of the underlying analysis. The four normal equations to determine the values of the parameters are:

$$\begin{aligned} a + .161b + .229c + .220d &= .383 \\ .161a + .161b + .075c + .049d &= .090 \\ .229a + .075b + .229c + .081d &= .122 \\ .220a + .049b + .081c + .220d &= .096\ . \end{aligned}$$

To ten places, Mathematica gives the solution:

a = .3152413832
b = .1634752755
c = .1540708799
d = .0279857542 .

These values then lead to:

A	B	C	From A, B, C	D	With D	With D′
0	0	0	.5657	a = .315241	.1783	.3874
0	0	1	.1193	a+d = .343227	.0409	.0784
0	1	0	.1023	a+c = .469312	.0480	.0543
0	1	1	.0517	a+c+d = .497298	.0257	.0260
1	0	0	.0663	a+b = .478717	.0317	.0346
1	0	1	.0197	a+b+d = .506702	.0100	.0097
1	1	0	.0457	a+b+c = .632788	.0289	.0168
1	1	1	.0293	a+b+c+d = .660773	.0194	.0099

The calculations are the same as we did in adding C to A and B. Here the first row tells us that the proportion of households which did not see any of A, B, and C was .5657; for these households, the proportion that saw D was .315241; therefore, ,31541(.5657) = .1783 was the proportion of households that saw D even though they did not see any of A, B, and C; finally, the last column gives .5657 - .1783 = .3874 as the proportion of households who also did not see D, granted that they had not seen A, B, or C. As a check on the arithmetic, the sum of the 6th column, "With D", = .3829 which, within a small rounding error, is the reach of D, .383. The resulting logical breakdown is:

		A		A′	
		B	B′	B	B′
C	D	.0194	.0100	.0257	.0409
	D′	.0099	.0097	.0260	.0784
C′	D	.0289	.0317	.0480	.1783
	D′	.0168	.0346	.0543	.3874

Note that there are sixteen entries in this table and that our original data only gave us 10 of the entries (reaches plus duplications). Thus, the Boolean analysis would have five unknowns to be solved for and this is sufficiently onerous to justify considering the alternative, namely to make four analyses each of which would provide an estimate of one of the unknowns (ABC, ABD, ACD, and BCD) and then a final analysis of ABCD, which would now have only one unknown, namely a = ABCD. In any case, this kind of difficulty increases geometrically as the number of media increase. Thus, for ten magazines - in fact, a modest number - there are 10 reaches and 45 duplications for a total of 55 known values whereas 1023 are needed to construct the complete logical breakdown. All of the needed entries are generated automatically by the multiple regression analysis so it is a worthwhile alternative to be considered when the number of magazines makes the Boolean analysis extremely onerous. Note, however, that there are also ways to simplify the Boolean analysis in such cases.

We now turn, in the next Section, to the application of Boolean methods to decision problems involving the selection of advertising media. There are a variety of such problems which seem "made to order" for Boolean analysis and which are not solvable by classical probability theory. Finally, any contingency table - a standard tool in many, perhaps most, scientific fields - generates important questions which are only answerable by Boolean methods. These will be the subject of Section 39, the last Section in this book except for some final remarks in Section 40.

Section 38. The Boolean Argument Applied to Media Data.

The advertising decision problems that result from media data are always of the same basic kind: there is drastically insufficient information. This is true even of magazine data, which is generally of greater quantity and quality than similar data for any of the other media: television, radio (AM and FM), newspapers, billboards, and so forth. Further, data on duplication between or among media is typically very sparse and of dubious quality. Nonetheless, decisions on advertising are made routinely and often involve very large sums of money. The advertising decision makers are no different than other decision makers in their desire to apply rational methods and to be able to justify their selections.

There are a variety of ways in which Boolean methods can be applied to improve the applicability of partial information to particular advertising decisions. Consider a common situation that arises frequently when an advertiser is attempting to reach a specialized audience. For example, high income professional households are difficult to reach and it often is the case that FM radio is one of the best media available for such households. It is frequently the case that the advertiser has partial information about the reach of FM radio to the desired households. Here is the kind of partial information that may be available. The advertiser knows the reaches of three magazines to the desired households: A is .2, B is .3, and C is .15. For these three magazines the advertiser knows the duplication of FM radio (designated M) and the magazines: MA = .06, MB = .09, and MC = .05. What is the estimated reach of FM radio (M)?

We can construct an equation to determine the reach of FM radio along the lines of the various examples shown in Section 31. Thus, we quickly can determine that

$$M = \sqrt[3]{\frac{(MA)*(MB)*(MC)}{A*B*C}}$$

For our data this gives M = .3107 as the best estimate of the reach of FM radio to the target households.

Let us switch our attention to the magazines alone. The commonest of all problems of incomplete information is having only partial information about the duplications (intersections) of magazines. For example, it is very common to know the reaches of the individual magazines and intersections by pairs and to have no explicit information about the intersections of more than two magazines. Further, it is the rule rather than the exception for there to be information on individual reaches, intersections by pairs of magazines, and intersections by three magazines at a time - and nothing further. Notice the inadequacy of such data. For an initial list of 15 magazines we would have 15 individual reaches, 105 intersections by pairs of magazines, and 455 intersections by triples of magazines - giving a total of 575 pieces of information. But complete information would require 32,767 pieces of information so what we have is only 1.8% of total information. If the initial list is of 25 magazines then we would have 25 individual reaches, 300 intersections by pairs, and 2,300 intersections by triples - or 2625 pieces of information out of a total of more than 33.5 million. Now, of course these comparisons are not intended to imply that we would ever need anything approaching such complete information. In fact, one of the major advantages of Boolean methods is that they enable us to calculate rather than to "look up". Section 40 will discuss this question at greater length.

Returning to the three magazines, suppose we know only the reaches, as given above, and the intersections by pairs: AB = .1, AC = .05, and BC = .08. If the advertiser is planning to use all three of these magazines then he or she would undoubtedly want an estimate of the intersection of all three, ABC. How can we get an estimate? First, and easiest, is to use a formula similar to the one we used for M above. Following along the lines of Section 31, it is easy to see that

$$ABC = \frac{(AB)*(AC)*(BC)}{A*B*C}$$

Here this gives us the estimate ABC = (.1*.05*.08)/(.2*.3*.15) = .0444. This might be called a "quick and easy" Boolean estimate of the intersection of three magazines. Suppose we knew the three reaches and the intersections AB and AC - but we do not know BC. What then? We would use

$$ABC = \frac{(AB)*(AC)}{A}$$

and find ABC = .05. Naturally, the less information is available for such formulas, the less confident we will be in our estimate. However, any such estimate is a lot better than guesswork.

What would be the standard Boolean estimate of ABC? This happens to be the simplest case of what we will see is the commonest use of Boolean estimates for media data. We developed a general formula in Section 31 that is always applicable in this common problem. We introduced a notation that is convenient here: let $\pi(n,i)$ be the product of all terms with n letters and with i of the letters negated. Thus, for our three magazines we would have:

$\pi(3,0) = ABC$
$\pi(3,1) = (ABC')*(AB'C)*(A'BC)$
$\pi(3,2) = (AB'C')*(A'BC')*(A'B'C)$
$\pi(3,3) = (A'B'C')$

In terms of our usual notation for the logical table these are

$\pi(3,0) = a$
$\pi(3,1) = e*b*c$
$\pi(3,2) = f*g*d$
$\pi(3,3) = h$.

Now, if you write out - for our three magazines - the formula

$$\frac{\pi(3,1)*\pi(3,3)}{\pi(3,0)*\pi(3,2)} = \frac{(ABC')*(AB'C)*(A'BC)*(A'B'C')}{(ABC)*(AB'C')*(A'BC')*(A'B'C)} = 1$$

can easily be evaluated and it will be found that the expression always equals one. The general formula, for arbitrary n, is a little harder to evaluate but it can be shown that the following formula always holds true for any n independent and random variables:

$$\frac{\pi(n,1)*\pi(n,3)*\pi(n,5)*\ldots}{\pi(n,0)*\pi(n,2)*\pi(n,4)*\ldots} = 1$$

Given n, we know that this equation holds true for any set of independent, random variables and we can solve it for any one unknown and be sure that this approach will always agree with the orthodox theory. The one that we almost overwhelmingly want to solve for - as we shall see - is $\pi(n,0)$, the intersection of all n magazines in the case of magazine data. Thus, for our example of n = 3 magazines we write

$$\pi(3,0) = a = \frac{\pi(3,1)\pi(3,3)}{\pi(3,2)}$$

However, we will not work with the right-hand side as written. Instead, we will use the letters we use for the logical table, expressing each of the other seven letters in terms of the unknown, "a". The result will be a cubic equation in "a". It is not a quartic because the a^4 cancels out on both sides of the equation.

It will be recognized that this is exactly Boole's analysis and we have found the other logical classes in terms of "a" in many of our examples throughout this book. In the present case - with our three magazines A, B, and C - we easily find

$e = .1 - a$	$f = .05 + a$
$b = .05 - a$	$g = .12 + a$
$c = .08 - a$	$d = .02 + a$
$h = .58 - a$	

b, c, e, and h are in the numerator and d, f, and g are in the denominator. Cross-multiplying by the denominator gives the equation in "a"

$$a(.02 + a)(.05 + a)(.12 + a) = (.05 - a)(.08 - a)(.1 - a)(.58 - a) \quad .$$

It is not necessary to work out the resulting cubic equation since the equation can be solved just as it is written, using numerical methods if desired. However, here is the cubic equation:

$$a^3 - .141a^2 + .01038a - .000232 = 0 \, .$$

The desired root is a = .0346564, which we will round off to a = .035. The resulting logical table is:

	A		A′	
	B	B′	B	B′
C	.035	.015	.045	.055
C′	.065	.085	.155	.545

Since we very often want to find the equations for b, c, . . . , g, h in terms of "a" it is worth showing how it can be done very quickly and on a computer. As an illustration of the procedure let us use the example with three magazines (D, E, and F) used at the end of Section 36. I will repeat the known information for convenience: D = .383, E = .334, F = .346, DE = .185, DF = .173, and EF = .158. The logical table is

	D		D′	
	E	E′	E	E′
F	a	b	c	d
F′	e	f	g	h

We begin with "a" as the unknown in its usual position. We know that the intersections are in each case of the same form: DE = .185 = a + e so e = .185 - a, similarly DF = .173 = a + b so b = .173 - a, and from EF = .158 = a + c we have c = .158 - a. This gives the following structure:

	D		D′	
	E	E′	E	E′
F	a	b = .173 - a	c = .158 - a	d
F′	e = .185 - a	f	g	h

We know that D = .383 = a + b + e + f so we can calculate f = .015 + a, similarly e = .334 = a + e + c + g so we can calculate g = -.009 + a, and from F = .346 = a + b + c + d we calculate d = .015 + a. This gives

	D		D′	
	E	E′	E	E′
F	a	b = .173 - a	c = .158 - a	d = .015 + a
F′	e = .185 - a	f = .025 + a	g = -.009 + a	h

Finally, since the sum of all eight entries must equal one we find h = .453 - a. We remember that a = bceh/dfg or adfg = bceh and we immediately have the equation

$$a(.015 + a)(.025 + a)(-.009 + a) = (.173 - a)(.158 - a)(.185 - a)(.453 - a)\ .$$

If you look at the end of Section 37-B you will find that this is exactly the same equation produced by an application of Jaynes' Principle to the same logical table.

In other words, in terms of applications to the logical tables of events (for example, household readership of magazines) Boole's method is exactly equivalent to the use of Jaynes' Principle. It becomes, then, almost a matter of choice as to whether we prefer to say "to apply Jaynes' Principle, use Boole's theory of probability" or "the strongest single argument in defense of Boole's method is that it is the way to maximize the entropy of the logical table, which is Jaynes' Principle". I say that it is "almost a matter of choice". In fact, I think that there are strong reasons to give major priority to Boole's theory at the same time that one recognizes that maximization of entropy is a strong argument in favor of Boole's theory. Boole's theory is more general, it works much more easily when there are several missing pieces of information, it provides a much easier way to find the equation that the two approaches have in common, and it has a unifying framework that is not evident in applications of Jaynes' Principle. Therefore, we are entitled to think of Boole's theory as the primary position but with Jaynes' Principle as a useful backup on occasion. At the end of this Section we will see an example where it is straightforward to apply Jaynes' Principle to obtain a result that is useful to the advertising decision maker but where the application of Boole's method would be extremely onerous.

Our previous example assumed that reaches of single magazines and intersections by pairs of magazines were available for a given set of three magazines. We then used the Boolean argument to find the logical table for all three magazines simultaneously. This meant that we needed only to estimate "a", the intersection of all three magazines. There are two obvious ways in which this example might be generalized. The first generalization arises if we know more information about the magazines under consideration. Thus, suppose we know in addition the intersections by triples of magazines. In this case if the advertiser is only considering three magazines then there is no need for Boolean calculations since the logical table for each set of three magazines is known from the information available. Suppose, however, that the advertiser is considering some set of four magazines. Then the logical table is not known since there is no information about the intersection of the four magazines. However, we can estimate this quantity, using the same general formula we used in the previous example.

To illustrate the procedure, suppose that we know the intersections for triples of magazines and we are considering the use of D, E, F, and G. The available information, then, is:

D = .383	DE = .185	DEF = .097
E = .334	DF = .173	DEG = .090
F = .346	DG = .154	DFG = .085
G = .310	EF = .158	EFG = .079
	EG = .151	
	FG = .136	

We need to estimate a = DEFG in order to determine the logical table for the four magazines. There will be sixteen entries in the logical table:

		D		D′	
		E	E′	E	E′
F	G	a	b	c	d
	G′	e	f	g	h
F′	G	j	k	l	m
	G′	n	p	q	r

From the triple magazine intersections we know, for example, that DEF = a + e = .097 so e = .097 - a; similarly, j = .090 - a, b = .085 - a, and c = .079 - a. Now, from the two-magazine intersections we know, for example, that DE = .185 = a + e + j + n and we now know e and j, whence n = -.002 + a. Similarly, f = -.009 + a, k = -.021 + a, g = -.018 + a, l = -.018 + a, and d = -.028 + a. Now, we have D = .383 = a + b + e + f + j + k + n + p, whence p = .143 - a. Similarly, q = .106 - a, h = .140 - a, and m = .123 - a. This gives fourteen of the entries. r = 1 - sum of these fourteen = .233 + a. Our formula for this case is

$$a = \pi(4,0) = \frac{\pi(4,1)*\pi(4,3)}{\pi(4,2)*\pi(4,4)}$$

Thus, we can write out the desired equation for "a":

$$a(-.002 + a)\ldots(-.028 + a)(.233 + a) = (.097 - a)(.090 - a)\ldots(.140 - a)(.123 - a)\,.$$

As written, this is an 8th degree equation in "a". If it were amalgamated into one polynomial = 0 then it would be of 7th degree but it would be a waste of time to do this since the equation is immediately solvable by numerical methods on computer just as it is written above. The root is a = .05345 and this is quite close to the actual value, .052, which I know in this instance but not generally. The logical table is:

		D		D′	
		E	E′	E	E′
F	G	.053	.032	.026	.025
	G′	.044	.044	.035	.087
F′	G	.037	.032	.035	.070
	G′	.051	.090	.053	.286

When one inspects a logical table that is this large or larger one cannot help wondering whether the advertising decision makers can discriminate sufficiently among these many classes so that such a detailed breakdown is useful. Thus, let us use again our notation with small letters, where deg means that exactly d and e and g were read, and not f . Then, can the decision maker make such refined discriminations that he or she can ascribe differing values to deg with .037 and def with .044? Probably, or perhaps, there is a difference in value to the advertiser from one typical household in one of these two classes as compared to a typical household in the other class. But who might have sufficient knowledge to make this discrimination? If such discriminations cannot be made then perhaps the decision maker needs the exposure distribution more than a logical table such as this one. Our present point, however, is that we have methods to develop a logical table if it is desired.

The second generalization arises if our data, for example, includes only intersections for pairs of magazines and we want to know the logical table for sets of four or more magazines. Suppose, to take a specific example, that we know only the reaches of individual magazines and the intersections of pairs of magazines and we are considering using D, E, F, and G - as in the preceding example. In this kind of situation there is no doubt that the most efficient approach is first to take the magazines three at a time and use standard procedures to estimate the intersection of each set of three (four of them) and then to use these estimates plus the known information to make the analysis of the preceding example. The only difference is that we are estimating the intersections of three magazines and then using these estimates in order to estimate the intersection for all four magazines. Thus, for DEF our estimate of the three-magazine intersection is DEF = a = .0996483 which we of course round off to .100 (actual is .097). For DFG our estimate is DFG = a = .080487 (actual is .085), for DEG our estimate is DEG = a = .093673 (actual is .090), and for EFG our estimate is EFG = a = .0802094 (actual is .079). It will be seen that in every case the estimate is reasonably close to the actual value. We now proceed exactly as for the preceding example except that we use these estimates of the three-magazine intersections instead of the actual values, as previously. We find that the root of the equation in "a" is a = .05386 which is very close to the .05345 that we found for "a" (= DEFG) when using the actual values.

In either kind of generalization we may have to face the question whether the decision maker has sufficient insight to make use of the logical table that we may construct. The logical table for n magazines will have 2^n cells. For five or six magazines it is surely reasonable to suppose that the logical table will be helpful for the decision maker but if we are dealing with 10 magazines and a table with more than 1000 cells or 20 magazines and a table with more than 1,000,000 cells then we surely ought to consider whether such a table will be of any use whatsoever for the decision maker. The chances are good that it will not be useful. This is particularly important because it will not be a simple computer run to generate such a logical table. Suppose, as is reasonably typical, that our data gives individual magazine reaches, intersections for pairs of magazines, and intersections for triples of magazines. Then in order to generate the target ten-magazine logical table we must make 210 runs to obtain estimates of four-magazine intersections, 252 runs for estimates of five-magazine intersections, 210 runs for estimates of six-magazine intersections, 120 runs for estimates of seven-magazine intersections, 45 runs for estimates of eight-magazine intersections, 10 runs for estimates of nine-magazine intersections, and, finally, one run to obtain the estimate of the ten-magazine intersection and the logical table. (All of this is merely an extension of what we did in the example of the preceding paragraph.) If you have reasonable software available these runs will be made very quickly and the whole process is completely feasible but if you have to write the software it can become quite time consuming.

Suppose our conclusion is that the logical table will not be helpful for the decision makers - and this will surely be true if n (the number of magazines) is large enough. (I suspect that n = 8 is already too large for a useful logical table.) This common situation ordinarily leads to the conclusion that the best help for the decision makers is the exposure distribution. The way this usually works is that the decision makers have already made preliminary selections of several (or many) sets of magazines and we want to use the computer to generate the resulting alternative exposure distributions. Often, in this context, the decision makers will want to know some specific intersections of small numbers of magazines and we can obtain these estimates by using the Boolean theory. The measurements of magazine reaches and intersections are already based on some congenial definition of the target audience. For example, advertisements for photographic film or for cameras may define the target audience as middle- to high-income families with at least one child of preschool age. Therefore, it is not unreasonable to ignore possible differences between the values of a typical reader in one logical class as compared to a typical reader in another class. It then follows that is surely sensible for the decision makers to compare alternative exposure distributions together with the costs in order to make a reasonably rational decision.

How can we determine the exposure distribution for a given set of magazines? This is a difficult question to answer. If we use the kind of Boolean analysis that we have been repeatedly illustrating in this book then there is no obvious alternative to making all of the numerous computer runs listed in the ten-magazine illustration of two paragraphs ago. It is clear that if we have that logical table of somewhat more than 1000 cells then we can find the exposure distribution directly by summing the entries in the table in accordance with the number of magazines represented by each entry. But it was already pointed out that a considerable amount of computation would be involved and the amount increases geometrically as the number of magazines in the specified set increases. For 20 magazines this approach would be out of the question.

Therefore, what we would like would be some reasonable approximation to the result that would be obtained from the logical table. There are many possibilities but I must put it on record that I have not found one that seems adequate. By the very nature of the Boolean data there are a considerable number of possible approximations. The difficulty is that usually the data are not known accurately enough to overcome the effects of rounding errors. Thus, the intersection of, say, five magazines is likely to be a fairly small number but the exposure distribution for ten magazines will have the probability of five exposures equal to the sum of 252 such intersections. As an illustration, a reasonably good approximation to the intersection of five magazines can be written as

$$(ABCDE) = \pi(3)*(\pi(1))^3/(\pi(2))^2$$

where $\pi(3)$ is the product of all of the intersections of three magazines from the five specified, $\pi(1)$ is the product of the reaches of the specified five magazines, and $\pi(2)$ is the product of all of the intersections of two magazines from the five specified. For one such intersection the actual value is .006 and the above approximation is 4.53E-24/1.685E-21 = .0027 and it is evident from the size of the numerator and denominator that the approximation is essentially noise - and, for that matter, so is the "actual". Another set of five magazines has the "actual" value = .015 and the approximation is 1.789E-20/1.868E-18 = .0096. Again, the magnitudes of the calculated values make it very doubtful whether the approximation is a decent one or not. For these and other related reasons I cannot recommend an adequate approximation to the exposure distribution that would result from the logical table of the lengthy calculations described two paragraphs ago.

Fortunately, we have an alternative approach. This is to apply Jaynes' Principle to the exposure distribution that would result from a given set of magazines. You will recall that applying Jaynes'

Principle means that if you have partial information about a probability distribution you should find the theoretical distribution that has the same stated characteristics as does the partially observed distribution and which maximizes the entropy, $-\Sigma p_i \log p_i$, where the logarithm can be used to any base but where the logarithms are frequently taken to the base 2. In this case the units of the entropy are called "bits". Then you use that theoretical distribution as if it were the actual observed distribution. There are a variety of reasons that make this approach a very sensible one. One of the most important is that this procedure results in the partial information having a lesser value than with any other distribution. In other words, the decision maker deliberately selects a theoretical distribution which downplays the important of the partial information to the maximum degree possible. This is a fine way to take a conservative stance.

An outstanding source book on Jaynes' Principle is Dr. Myron Tribus: Rational Descriptions, Decisions, and Designs; Pergamon; 1969. In this book Dr. Tribus gives many fascinating applications of the principle and presents a number of important theorems concerning the use of the principle. In our use of the principle we will be working with the exposure distribution. Suppose there are 4 magazines under consideration. The exposure distribution will look like this:

Exposures	Probabilities
0	p_0
1	p_1
2	p_2
3	p_3
4	p_4
	1

The partial information we will take account of provides three restrictions on the p's:

$$1.\ \sum_0^4 p_i = 1$$

$$2.\ \sum_0^4 ip_i = S_1$$

$$3.\ \sum_0^4 i^2 p_i = S_1 + 2S_2$$

You will recall that assuming this knowledge is equivalent to assuming that we know the mean and variance of the exposure distribution and we are going to find the theoretical distribution with the same mean and variance as observed and which maximizes the entropy.

The standard mathematical method for finding a constrained maximum as here (meaning a maximum of

$$-\sum_0^4 p_i \log p_i$$

subject to the constraints given by the three equations) is to use Lagrangian multipliers. There is one for each restricting equation and they are often designated as λ_0, λ_1, and λ_3. Tribus shows that the maximum entropy distribution is given by

$$p_i = e^{-\lambda_0 + \lambda_1 i - \lambda_2 i^2}$$

Here are some examples. For magazines A, B, C, D, we have S_1 = .993 and S_2 = .513. The maximum entropy distribution is p_i = e^(-.934326+.0218943i-.196417i²), giving

Exposures	Probability	Actual
0	.392851	.393
1	.329939	.336
2	.187083	.168
3	.071619	.091
4	.018510	.012
	1.000002	

The reader can verify that the other two restrictions are exactly met. For magazines D, E, F, G, S_1 = 1.373 and S_2 = .957. The maximum entropy distribution is p_i = e^(-1.25203 +.18993i - .14982i²), giving

Exposures	Probability	Actual
0	.285924	.285
1	.297625	.304
2	.229592	.216
3	.131254	.143
4	.055608	.052
	1.000003	

Another set of four magazines from the same data base is A, D, G, H, with S_1 = .944 and S_2 = .385. The maximum entropy distribution is po = e^(-.991 + .360998i - .342111i²), giving

Exposures	Probability	Actual
0	.370834	.388
1	.377905	.334
2	.194281	.223
3	.050388	.054
4	.006593	.001
	1.000001	

As a larger example, the data base from which I have been taking these four-magazine examples has a total of eight magazines (A - H) with S_1 = 2.073 and S_2 = 7.023. The maximum entropy distribution is p_i = e^(-1.64063 + .260833i - .102767i²), giving

Exposures	Probability
0	.193858
1	.227055
2	.216529
3	.168127
4	.106291
5	.054714
6	.022932
7	.007825
8	.002174

For this larger distribution I do not know the actual values. I have never seen data that gave more than the intersections (or reaches) of sets of five magazines at a time and these data are marvelously expensive and surprisingly voluminous. Of course, one has no reason to be surprised at the size of the data. These data were purchased by a large advertising agency and the information was collected for each of nine family size-income classes: high, middle, and low incomes for households with no children, with at least one child of pre-school age, and children, but none of pre-school age. There were 23 magazines included in the data. This means that for each of nine classes there had to be given 23 reaches, 253 intersections of pairs, 1771 intersections of triples, 8855 intersections of four magazines each, and 33,649 intersections by five magazines. It would take a very large computer-printed book to contain the information for any one of the nine classes. It is because of such very strong limitations on the availability of information that we are generally forced to calculate estimates of desired information rather than to look them up.

The use of Jaynes' Principle on the exposure distribution is a very valuable addition to the tools that are available to help advertising decision makers. It will be remembered that when both Boole's theory and Jaynes' Principle can be applied to the same logical table they turn out to be equivalent to one another. This is certainly not a proof of anything at all but perhaps it provides a basis for believing that the use of Jaynes' Principle on the exposure distribution will be at least a rough-order approximation of what we might obtain from Boole's procedure if we were willing and able to undertake the calculations.

Calculating the maximum entropy distribution is not mathematically difficult but it certainly requires more mathematical knowledge than the other parts of this book. Probably, persons who want to make the necessary calculations should have a program that would perform the calculations. In that case it will be quite easy to find the desired distribution. Rather than present the somewhat technical material here, I have included Appendix D that outlines the procedure to find the maximum entropy distribution.

Section 39. The Structure of Contingency Tables.

Cross-classified tables of frequency counts are often called contingency tables. Such tables can have any number of dimensions, where the number of dimensions is the number of categorical variables according to which the population is cross-classified. The use of chi-square to analyze two-dimensional contingency tables has long been one of the main topics of basic courses in statistics. For about sixty-five years the development of methods for the analysis of contingency tables of three or more dimensions has been a major effort in statistical theory.[2] The literature on the analysis of such multi-dimensional tables has had major emphases on loglinear models, additive models, and models based on partitioning chi-square. With the ready availability of computers there has been a great increase in the number of books and articles devoted to this subject and, in particular, to the use of loglinear models.

Much of this analysis is intended to uncover what may be called the structure of the contingency table. The simplest structure would be one where the dimensions jointly behave like independent random variables. Thus, in a three-dimensional table the probability in each cell of the table would be the product of the corresponding probabilities of the three categorical variables on which the dimensions are based. A divergence from independence is called interaction and the goal of the analysis might be summarized as being the discovery of the minimum number of interactions necessary to explain the table. The operational definition of "explain" here is that the table is explained if it can be reconstructed, given only the basic dimensional probabilities and the necessary interactions.

The point of these preliminary remarks is that Boole's theory is an alternative way to undertake the analysis of the structure of contingency tables. Since the analysis is based on Boolean logical classes it follows that the contingency table must be based only on dichotomies but this is not nearly so severe a restriction as it might seem. Often, the dimensions with more than two classes can be collapsed into a dichotomy with little loss for the analysis. In any case, there are significant benefits associated with using Boole's analysis in this way. The structure of the argument is quite straightforward and this may enhance its appeal for some analysts.

Suppose we are considering using three magazines for advertising. The typical hypothesis that we will want to test is: given the single magazine reaches and the intersections by pairs of magazines we can calculate, within ordinary sampling fluctuations, the intersection of all three magazines. We need a reasonable, and conservative, test of significance. There are two major methods that might be used. One is to use chi-square and compare the calculated frequency with the actual frequency. This, of course, would require that we have the data in the form of frequency counts rather than probabilities. The other is to use the normal approximation to the binomial and calculate the probability of obtaining by chance an observed value as far or further from the predicted value. The two tests are closely equivalent. Since all of our examples in this book have been based on probabilities, we will use the second approach.

Here is some data on male readership of mass magazines:

	A		A′	
	B	B′	B	B′
C	.0808	.0696	.0775	.1653
C′	.0603	.1400	.0601	.3464

It is obvious that these magazines are not independent by pairs. Thus, p(A) = .3507 and p(B) = .2786 but p(AB) = .1411, almost half again as large as the .0977 that would be expected if the two magazines were independent. This granted, a reasonable question that might be asked is: Does the dependence by pairs of magazines exhaust the whole dependence of these data or is there still more that is not accounted for by the pairs?

To answer this question we can assume that we are given single magazine probabilities and the intersections by pairs and then use Boole's analysis to predict the triplication. If the prediction is good, meaning that the fit between the predicted values and the actual values is satisfactory then we can surely conclude that the dependence by pairs is all the dependence displayed by the data. We intend to use our standard Boolean approach to construct a prediction for "a" = ABC, which we will compare with the observed .0808. We can write all the logical equations in terms of one unknown, "a", in our usual way. We find:

$$
\begin{aligned}
e &= .1411 - a\\
b &= .1504 - a\\
c &= .1583 - a\\
f &= .0592 + a\\
g &= -.0207 + a\\
d &= .0845 + a\\
h &= .4272 - a
\end{aligned}
$$

Then, as usual, we write out the polynomial equation adfg = bceh and find the root a = .080829, very nearly identical to the observed value, a = .080796. This is so close that no test of significance is needed but we will nonetheless illustrate the argument and the procedure. We use the predicted value (the root of the polynomial equation) as the expected value of "a". We need the standard deviation of the normal distribution of observed values and this is given by $\sqrt{[p*(1-p)/n]}$, where n is the sample size and p is the value predicted for "a". For these data n = 4827 so the standard deviation is .0039. We would calculate (.080829-.080796)/.0039 and this is trivially small so we have no reason to reject our hypothesis. Our conclusion is that knowing single magazine reaches and intersections of pairs of magazines enables us to calculate the intersections of triples. For these data there is no indication of interaction additional to that of pairs of magazines.

To show that this approach can lead to results of interest for magazine data, consider adding a magazine to the above three. When we add magazine D and attempt to predict the three-magazine and four-magazine intersections from the singles and intersections by pairs we find the fit is extremely poor. Here are some examples from the complete table:

Magazines	Actual	Predicted	Standard error
ABCD	.0433	.0354	.00266
ABCD′	.0157	.0261	.00229
A′BCD	.0445	.0342	.00262

The normal deviates are, respectively, 2.97, 4.54, and 3.93. The first difference is highly significant and the other two are very highly significant. We can safely conclude that there is interaction among these magazines that is additional to that by pairs. We can then ask whether all of the additional dependence is exhausted by the interactions among the magazines taken three at a time. Using Boole's analysis we can estimate ABCD from the singles, the intersections by pairs, and the intersections by triples. We find the estimated ABCD is a = .03837 and the normal deviate is 1.78, not significant. We conclude that the dependence among these magazines appears to be exhausted by taking the intersections by pairs and by triples.

I want to make it abundantly clear that this kind of application of Boole's theory is not in any way confined to the media data used in advertising. The contrary is the case. Boole's theory can be applied to any dichotomous contingency table. In order to show this, I have selected some data from a book by Stephen E. Fienberg.[3] Thus, he quotes some counts for structural habitat categories for 192 *Anolis* lizards of Bimini.[4] Let A = perch diameter is ≤ 2.5 inches; B = perch height is ≤ 5 feet; C = *sagrei* adult males and C′ = *augusticeps* adult males. The data are:

	A		A′	
	B	B′	B	B′
C	.25	.00781	.43750	.09375
C′	.01562	.10938	.01042	.00521

These data are not particularly prepossessing and they are clearly dependent by pairs. Is there dependence in addition to that by pairs of categories? Using Boole's method, we estimate a = ABC to be .24375, using the singles and the intersections by pairs. This is obviously an excellent fit so we do not need to calculate the normal deviate. (It is only about .2.) We conclude that there is no dependence among these data beyond that by pairs of categories.

Another example shows that it is often straightforward to collapse the data to make the contingency table dichotomous. Fienberg gives data dealing with the respiratory condition of English coal miners of various ages.[5] Two of the dimensions are dichotomous: A = breathlessness and B = wheeze. The third dimension is age, in nine five-year classes. Define C = < 45 years old and the resulting contingency table is dichotomous:

	A		A′	
	B	B′	B	B′
C	.02057	.00749	.04961	.51925
C′	.07937	.02533	.05065	.24773

For these data n = 18,282. We follow the usual procedure to obtain a quartic equation in "a" (a cubic after one subtraction). The root is a = .01822. The standard error is very nearly .001 and the normal deviate is 2.38, with probability .0173, which is certainly significant. We conclude that the dependence among these data is not exhausted by the dependence by pairs.

This kind of lumping can equally well be used when a dimension is not necessarily based on measurement data. Thus, Feinberg gives data on a consumer blind test.[6] Here there are four dimensions and three of them are dichotomous. The fourth dimension is water softness, divided into soft, medium, and hard. Treat this dimension as soft and not-soft and all four dimensions are dichotomous. For these data n = 1008. The actual value of ABCD is a = .018849. Following our usual procedure and using only the singles and the intersections of pairs the predicted value for ABCD is .020436. The standard error is .004456 and the normal deviate is less than one. Therefore, we can accept the hypothesis that all of the dependence in these data is exhausted by taking the dependence by pairs of categories.

The analysis is fully as effective on tables with more dimensions although the calculations become sufficiently onerous to require doing them on a computer - preferably with a program written for this specific purpose. Fienberg gives a five-dimensional table, with each dimension being dichotomous.[7] The dimensions here depend on whether the individual subjects: read newspapers, listen to the radio, do "solid" reading, attend lectures, and have good or poor knowledge concerning cancer. Using Boole's method, we find that the estimates based only on singles and intersections by pairs are very poor. However, when the triplications are included the predictions become quite good. In particular, the observed values of the quadruplications do not differ significantly from the predictions. Therefore, we are entitled to conclude that the dependence in these data is exhausted by taking the dimensions two and three at a time.

It should be noted that it is not at all necessary to take all of the duplications or all of the triplications and so forth, as has been done in each of the examples given above. Using Boole's method it is straightforward to investigate, for example, whether some specified duplications and some specified triplications, say, are sufficient to predict the whole table. No example of this was given because this approach involves more specialized knowledge of the nature of the dimensions involved in the contingency table. Therefore, such an analysis does not seem appropriate here.

It seems to me that this is an attractive and useful method of uncovering the structure of contingency tables. I am not claiming that this approach provides means of accomplishing analytical purposes that cannot otherwise be achieved although it is unlikely that other methods will produce exactly the same results. However, Boole's method is at least a good alternative approach to this important class of problems.

Finally, it is worthy of emphasis that the methods of this Section and the preceding one are exactly Boole's analysis. There is no change from his argument and there are no errors in it. The point of this comment is that the supposed error in Boole's argument, discussed previously, does not in any way affect the procedures discussed here.

Section 40. Boole's Theory of Probabilities.

In Chapter I in my discussion of the Challenge Problem I said that George Boole had bad luck in his choice. This is surely true. There are only two seeming difficulties with the Boolean theory of probabilities that he did not himself resolve. That the Challenge Problem should include both of them is extremely bad luck. It is my opinion that if Boole had selected a different problem, without the two difficulties, there might well have been a completely different response to his theory.

In any case, I have attempted to show in this book that the problems can be dealt with although it is certainly true that there is no general theory encompassing Boole's that can handle the problems. I am convinced that if there were such a general theory, George Boole would have found it. So, in the absence of such a general theory, I have argued differently. I have stated perhaps too many times that only a tiny proportion of real-world problems would incorporate either one of the difficulties. The great majority of real-world applications - and there are many - would involve nothing but straightforward Boolean analyses of exactly the kind George Boole gave us.

Boole was a kindly and modest man. His life was foreshortened - in the beginning by poverty, in the end by his untimely death. I believe that the great conception he had and presented to us in his "An Investigation of the Laws of Thought" was equaled in human intellectual history only by those of Aristotle and Leibniz. It seems to me that he deserved far better from his posterity than he received. He succeeded fully in his goal and no one seems to know it. If anyone happens to read this book and as a result shares my own enthusiasm for the remarkable intellectual structure that George Boole created then this author's modest efforts will be richly rewarded.

APPENDIX A

BOOLEAN ALGEBRA

A BRIEF SUMMARY

There are two general classes in Boolean algebra:

1 = the class that includes everything (every class is included in 1)

0 = the null class (the class with no members)

The names of these two classes are 1 and 0. There is no implication that arithmetic will be relevant to class manipulations but the reason these two classes are named 1 and 0 is that these classes behave within Boolean algebra in very much the way that the numbers 1 and 0 behave in arithmetic. We will see this shortly.

A = the class of things with characteristic A

1 - A = the class of things that do not have characteristic A. This class is called the negation of class A. Boole often writes the negation of A as

$A' = 1 - A$ = the negation of class A.

Notice that the negation of negation returns to the original class:

$A' = (1 - A)$ and $(1 - A)' = (1 - (1 - A)) = A$.

In Boolean algebra there are three operations that can be performed on two or more classes:

Addition of classes, as A + B, where

A + B = all those things that are included in A or included in B, or are included in both A and B

Multiplication of classes, as AB , sometimes written A*B, where

AB = A*B = all those things that are included in both A and B

Implication, as $A \subset B$, meaning that membership in class A implies membership in B

$A \subset B$ is defined as being equivalent to $A*(1 - B) = A*B' = 0$.

In our discussion of Boole's theory of probability we will not need implication.

We have immediately that $A0 = 0$ and $A + 0 = A$. These are similar to the way 0 operates as a number in arithmetic. Generally, there is no operation of subtraction in Boolean algebra other than the subtraction required in the negation of a class, namely that 1 - A is the negation of A. Boole occasionally used subtraction but the subsequent development of his algebra has eliminated subtraction (and division). The reason is mostly that we could only meaningfully subtract A, say, from B if A is a subset of B and we do not ordinarily know this. Notice that an important conclusion from the lack of subtraction is that the single equation, $A + B = 0$ is automatically equivalent to the two equations, $A = 0$ and $B = 0$. This obviously generalizes. When a sum of any number of Boolean terms = 0, then each individual term, taken separately, must be equal to 0. Further, $A*1 = A$ but $A + 1 = 1$. The first of these two is similar to the way the number 1 operates in arithmetic but the second is not. This is one of the unique characteristics of Boolean algebra. Since there is no division in Boolean algebra, the equation $A*B = 1$ is equivalent to the two equations, $A = 1$ and $B = 1$.

To illustrate the basic ideas a little further, note that $A + A' = 1$ and $A*A' = 0$. The unique character is this algebra is shown by the pair of equations, $A + A = A$ and $A*A = A$. This is in sharp contrast to the more usual kind of algebra. However, most of the manipulations of ordinary algebra carry over unchanged to Boolean algebra. Specifically,

$$A*(B + C) = A*B + A*C$$

and

$$(A + B) + C = A + (B + C).$$

The major additional manipulations we will need are related to negating classes. Given A we write the negation of A as A′ (or as (1 - A)). But what is the negation, say, of $(A' + B)$? We can write the desired expression just as easily as for A alone. The negation of $(A' + B)$ can be written $(A' + B)'$ but the question is: What is this negation in terms of A and B? The answer to any such question is given by two theorems called De Morgan's theorems:

$(A*B)' = A' + B'$ and $(A + B)' = A'* B'$.

These two theorems apply to any number of terms. The rule is: To negate an expression that connects a series of terms by one operation (addition or multiplication), negate each of the terms and change the operation (from addition to multiplication and vice versa). Thus, $(A' + B)' = A*B'$. As another example: $(A'*B*C*D')' = A + B' + C' + D$. Often, one must go through the procedure several times in order to obtain the final answer. Thus, $(AB' + A'C)'$ = first; $(AB')'*(A'C)'$ = second; $(A' + B)*(A + C')$ = third;

$A'C' + AB + BC'$ (since $A'A = 0$).

We need one further, highly important, concept: the normal form. In Boolean algebra, any series of terms involving X and X′ is called a Boolean function of X. Such a function is said to be in normal form when it is written thus:

$f(X) = AX + BX'$

Here A and B are independent of X (meaning that they do not include either X or X′. Notice that this can always be accomplished. One of the ways in which Boolean algebra differs from ordinary algebra is that we can always very easily add any new variable to any Boolean equation. Suppose, then, that we have the Boolean expression $AB + A'X'$. This function has one term without either X or X′. We want the normal form in terms of X. This means that we must make every term in the expression have either X or X′. We can do this easily. Remember that, for example, $X + X' = 1$. We can multiply any term by 1 without changing its value. Therefore, we write

$$AB' + A'X' = AB'(X + X') + A'X' = AB'X + (AB' + A')X' = f(X)$$

and we have the desired normal form.

It is very easy to write down the normal form. Suppose that X is the variable of interest and that there is an equation in X which is equal to 0. Using ordinary functional notation we can write the equation as f(X) = 0. In Boolean algebra, as in ordinary algebra, we can substitute values for the variable, X. However, for Boolean algebra the only two values for variables are 0 and 1. Let f(1) be the value of f(X) when X = 1 and let f(0) be the value of F(X) when X = 0. Then it can be shown that the normal form for the equation f(X) = 0 is

$$f(1)X + f(0)X' = 0 .$$

Thus, take our immediately preceding example and write f(X) = AB′ + A′X′ = 0. Substituting X = 1 (meaning also that X′ = 0) gives f(1) = AB′ and substituting X = 0 (meaning also that X′ = 1) gives

f(0) = AB′ + A′. Therefore, the normal form is

$$f(X) = AB'X + (AB' + A')X' = 0 ,$$

as before.

There are several uses for the normal form. The one we are most interested in is that the normal form gives us a particularly easy way to eliminate an unwanted variable from a Boolean equation. Suppose we want to eliminate X from the equation f(X) = 0. This can be accomplished immediately by writing

$$f(1)*f(0) = 0 .$$

By inspection of the normal form it is apparent that there will be no X nor X′ in this equation. In order to see this, return to the normal form as written in terms of f(1) and f(0):

$$f(1)X + f(0)X' = 0.$$

As we stated earlier, when a sum of several terms in Boolean algebra is equal to 0 then each term take separately is equal to 0. Thus, from the preceding equation it follows that f(1)X and f(0)X′ are each separately equal to 0. But either X = 0 and X′ = 1 or else X = 1 and X′ = 0. In the former case it follows that f(0) must be 0 (to make f(0)X′ = 0) and in the latter case it follows that f(1) must be 0 (to make f(1)X = 0). Therefore, in any case either f(1) or f(0) must equal 0 so f(1)f(0) = 0 always. This is the equation without any X. In our example we have f(1)*f(0) = AB′ = 0 and this is the equation with X eliminated.

It is worthy of mention that the elimination of variables is a particularly important step in Boole's theory of probability.

APPENDIX B

AN EXAMPLE OF THE COMPLEXITY OF BOOLE'S METHOD

Most of Boole's examples are surprisingly complicated. Merely attempting to verify one of the steps in a typical example of Boole's is very likely to require a dismaying amount of algebra and to try to work out the whole solution can only be described as being remarkably difficult. Boole himself at one point described the necessary calculations as being "repulsive" and this author surely concurs. This is really unexpected since there is no "higher" mathematics involved. All that is ever needed is algebra but the necessary manipulations are almost always very lengthy and usually subject to all kinds of notational errors.

This is so much the case that I seriously doubt whether anyone has worked out an example using Boole's method since the brief flourishing of this kind of analysis - an algebra of logic and probability - in the last quarter of the nineteenth century. This flourishing was associated with persons such as Hugh MacColl, Alexander Macfarlane, Christine Ladd-Franklin, and Charles Sanders Peirce. Some of them worked through Boole's method if for no other reason than that they wanted to argue that Boole's theory was inferior to one of their's. After modern symbolic logic was launched, there was no further interest in the whole approach of this brief "school". It is certainly true that no one has shown any knowledge of how to apply Boole's theory since roughly 1900 - certainly not Keynes. Hailperin, on the other hand, in his magisterial book, Boole's Logic and Probability, proved that Boole's system was coherent and put it on a sound mathematical foundation. He also solved Boole's general problem of probability in the sense of giving upper and lower bounds for the probability desired. However, this was not Boole's sense and is not in any way Boole's solution (see Section 6). Therefore, it is fair to say that Hailperin also did not apply Boole's theory.

Therefore, it seems appropriate to demonstrate the kind of complexity that is usually encountered in Boole's examples and that is the purpose of this Appendix. We will use Boole's Example 2 of Chapter XVIII, "Elementary Illustrations of the General Method in Probabilities", p. 278. Occasionally, Boole gives examples in which the known information and the desired unknown probability are considerably more complex than would ever be likely to occur in a realistic problem. His motivation for such examples seems to have been that he wanted to show that his method would produce a correct answer in cases where the unaided mind would have difficulties because of the unfamiliarity of the data. This example seems to be of this kind.

Here is the problem, as stated by Boole:

> "The probability that one or both of two events happen is p, that one or both of them fail is q. What is the probability that only one of them happens?"

Boole lets x and y represent the respective events. The data, then, are

$$\text{Prob.}[xy + x(1\text{-}y) + (1\text{-}x)y] = p; \quad \text{Prob.}[x(1\text{-}y) + (1\text{-}x)y + (1\text{-}x)(1\text{-}y)] = q$$

and the desideratum is $\text{Prob.}[x(1\text{-}y) + (1\text{-}x)y]$.

The elementary events here are x and y and we do not know the probabilities of x and y. The two events for which we know the probabilities and the event for which we want to determine the probability are all compound events, defined in terms of the elementary events, x and y. One of Boole's most basic and important arguments is that each compound event should be given a designation by a single variable and that there is no important distinction between elementary events (sometimes called "simple") and any compound events. Therefore, Boole designates the compound events thus:

$$xy + x(1-y) + (1-x)y = s, \qquad x(1-y) + (1-x)y + (1-x)(1-y) = t, \qquad x(1-y) + (1-x)y = w.$$

Since we do not know the probabilities of x and y, we cannot deduce the unknown probability of w from the above equation for w in terms of x and y. Our first goal, then, is to express w as a function of s and t. Boole's method for doing this is to convert each of the three definitions above into an equation equal to zero. Then all three of these left-hand sides equal to zero can be added and we will have one large equation equal to zero. Our procedure for converting any one of these three equations into an equation equal to zero is a standard one in Boolean algebra: if we have an equation, say $A = B$, then multiplying by B' gives $B'A = 0$ (since $B'B = 0$) and multiplying by A' gives $0 = A'B$ (since $A'A = 0$). Therefore, the equation $A = B$ is equivalent to the equation $A'B + AB' = 0$. The rule, then, is simple: Negate the right-hand side and multiply by the left-hand side; negate the left-hand side and multiply by the right-hand side; add these two quantities together and set the sum equal to zero. Then add all three of the resulting equations together to obtain one left-hand side equal to zero. The difficulty in so doing is only obtaining the negations of the left-hand sides. Note that it is quite important in this step to write each negation in the form (1-x) rather than as x'. This is because in some subsequent substitutions of values it is very much easier to avoid mistakes when (1-x) is written rather than x'.

Using De Morgan's theorem as given in Appendix A, we find, for example, that the negation of the left-hand side of the equation for s is (1-x)(1-y) so the first equation will become

$$s(1-x)(1-y) + (1-s)[xy + x(1-y) + (1-x)y] = 0 .$$

Proceeding similarly, and summing the left-hand sides, we get:

$$s(1-x)(1-y) + (1-s)[xy + x(1-y) + (1-x)y] + txy + (1-t)[x(1-y) + (1-x)y + (1-x)(1-y)] +$$

$$+ w[xy + (1-x)(1-y)] + (1-w)[x(1-y) + (1-x)y] = 0 .$$

Now Boole wishes to eliminate x and y from this equation since we do not know the probabilities of x and y. Consider how we could proceed to eliminate x only. Treat the whole left-hand side as f(x) and, in effect, find the normal form for f(x). This means substitute $x = 1$, thus finding f(1), and then substitute $x = 0$, thus finding f(0). As shown at the end of Appendix A, $f(1)*f(0) = 0$ will be the original equation with x eliminated.

When there are, as here, two variables to be eliminated this can be done in one step. Boole develops the general procedure for any number of variables to be eliminated. For two we simply consider the left-hand side to be f(x,y) and obtain the four quantities: f(1,1), f(1,0), f(0,1), and f(0,0). The desired equation, with x and y both eliminated, will be the product of these four quantities set equal to zero. Proceeding thus, we find:

$$f(1,1) = 1 - s + t + w$$

$$f(1,0) = 3 - s - t - w$$

$$f(0,1) = 3 - s - t - w$$

$f(0,0) = 1 + s - t + w$

These must be multiplied together and the product set equal to zero. Doing it in two steps, I find

$f(1,1)*f(0,0) = 1 - s - t + 3w + 2st$ and $f(1,0)*f(0,1) = 9 - 5s - 5t - 5w + 2st + 2sw + 2tw$.

Multiplying these two products together gives an equation that expresses w in terms of s and t:

$9 - 9s - 9t + 7w + 10st - 4sw - 4tw = 0.$

We now want to solve this equation for w, obtaining

$$w = \frac{-9 + 9s + 9t - 10st}{7 - 4s - 4t}$$

This expresses was a function of s and t, for which we know the probabilities. Boole wants to expand this expression for was a function of all possible terms involving s and t, which he calls the constituents of the expansion. The possible terms are st, s(1- t), (1- s)t, and (1- s)(1- t). Boole calls this developing was a function of s and t. We need the coefficients of the four possible constituents and Boole proves that there are four possible coefficients: 1, 0, 0/0, and all other possible numbers, which Boole designates by 1/0. Boole writes this in a general form as:

$$w = A + 0B + \frac{0}{0}C + \frac{1}{0}D$$

Boole defines his key quantity as V = A + B + C. As we will see shortly, each of the quantities in V can be directly interpreted in terms of the meaning of our solution for the probability of w. (Note that D plays no further role in our argument.)

It is straightforward to find the coefficients of the constituents, using the equation for was a fraction involving s and t. Let that fraction be designated $\phi(s,t)$. Then the coefficient of st is $\phi(1,1)$, the coefficient of s(1- t) is $\phi(1,0)$, the coefficient of (1- s)t is $\phi(0,1)$, and the coefficient of (1- s)(1- t) is $\phi(0,0)$. We find $\phi(1,1) = -1/-1 = 1$; $\phi(1,0) = 0/3 = 0$; $\phi(0,1) = 0/3 = 0$; and $\phi(0,0) = -9/7 = 1/0$. Therefore, we have shown that

$$w = st + 0s(1- t) + 0(1- s)t + \frac{1}{0}(1- s)(1- t)$$

Here A = st, C = 0, and V = st + s(1- t) + (1- s)t = s + (1- s)t by algebraic simplification. Boole now defines V_s = the aggregate of the constituents in V which contain s as a factor - here $V_s = s$ - and V_t similarly as all of the constituents in V which contain t as a factor - here $V_t = t$. (in following these definitions of V_s and V_t one must think of V written thus: V = st +st′ +s′t. Then it is clear that $V_s = st + st' = s(t + t') = s$ and, similarly, $V_t = st + s't = t(s + s') = t$.)

We have been following Boole's presentation of his "General Rule" in Paragraph 17 of Chapter XVII. We are now ready to take the step that Boole usually refers to as "passing from Logic to Algebra" or some equivalent phrase. The general form of the resulting equations is

$$\frac{V_s}{p} = \frac{V_t}{q} = \frac{A + zC}{w} = V$$

Here z is a symbol meaning that some unknown part, from none to all, of any logical class in C will be included in w. This possibility does not arise in this example because C = 0. The specific equations that we need to solve, then, are

$$\frac{s}{p} = \frac{t}{q} = \frac{st}{w} = s + (1-s)t$$

By "solving" we mean merely to eliminate s and t in the expression for w. Even these equations are not quite as easy to solve as they appear. I found s = (p + q - 1)/q and t = (p + q - 1)/p and then substituted these values in the right-hand equation above to find w = p + q - 1, as given by Boole on p. 278.

Three comments are well worth adding. First, it is easy to see that the answer is correct. This can be seen in two ways. The first way requires noticing that as the three compound events are defined we have immediately that w = s + t - 1. By Boole's assumptions the compound events are treated as if they are independent so immediately we would have w = p + q - 1. However, if we noticed this relationship then we could have avoided the somewhat laborious algebra involved in finding the product of the coefficients of the constituents, as above. Instead, we work directly with w = s + t - 1 and expand w on the basis of this equation. Thus,

s	t	w
1	1	1
1	0	0
0	1	0
0	0	-1 or 1/0 .

We get the same expansion as by the more difficult manipulations given above. V remains exactly the same and all the rest of the argument follows.

Second, we can solve the Boole's probability problem very much more easily by a different kind of standard approach. A tabular representation of this problem looks like this:

	X	X'
Y	a	b
Y'	c	d

where the entries in the body of the table are unknown probabilities. Taking the information assumed by Boole we can write three equations:

a + b + c = p
b + c + d = q
a + b + c + d = 1.

We are asked to find b + c. It is immediately obvious that (a + b + c) + (b + c + d) - (a + b + c + d) = b + c so b + c = p + q - 1. As will be shown in this book, the tabular approach enormously simplifies almost all of Boole's examples.

The third comment is analogous to a well-known evaluation by Hardy and Littlewood of the 25 or so new theorems announced by Ramanujan in his initial letter to Hardy. Hardy was very much impressed by the novelty of these theorems and he asked Littlewood to meet him at the faculty club so the two of them - perhaps the third and fifth most powerful living mathematicians at that time - could consider these theorems. So doing, they both concluded that they had never seen anything like at least fifteen of the stated results. This was the occasion of their initial evaluation, namely, that they believed that the theorems must be true because it would have been impossible for anyone to conjure them up if they were false. I think that the same principle can be applied to Boole's system. When such lengthy and complicated manipulations lead to the right answer repeatedly in unusual problems in probability then one is entitled to conclude that the system **must** be basically correct. I believe that Boole constructed quite a few of his examples with this idea in the background. Of course, this does **not** mean that Boole's system is correct in every detail. It is not. There is at least one error and we will discuss it subsequently. This error rarely arises in practical problems so it is fair to summarize to the effect that the basic system is sound and will definitely support some important applications that cannot otherwise be accomplished.

APPENDIX C

ON SOLVING THE BOOLEAN EQUATIONS

The solution of the Boolean equations obtained by the procedures outlined in the text is straightforward. We may take the Challenge Problem as summarized in Section 19 as an example. It will be recalled that we will have equations relating a, b, . . . , h of the logical table to the known probabilities. The Boolean variables for the Challenge Problem are t = AC and u = BC plus A and B. C is not a variable because we do not know the probability of C.

By ordinary logical multiplication we get the six equations, given in Section 19, of the form: tuAB = a, tu'AB' = b, . . . , t'u'A'B' = d + h. Dividing each left-hand side by the right-hand side, equating the results, and setting them all equal to Boole's V gives the Boolean equations:

$$\frac{tuAB}{a} = \frac{tu'AB'}{b} = \frac{t'uA'B}{c} = \frac{t'u'AB}{e} = \frac{t'u'AB'}{f} = \frac{t'u'A'B}{g} = \frac{t'u'A'B'}{d+h} = V$$

We do not need to use the relationship with V in solving the equations.

The easiest way to solve the equations is to make some simple changes in the variables. There is no need to introduce new variables since the changes we will make are all easy ones. First, divide all seven numerators by tuAB and put t = t'/t, u = u'/u, A = A'/A, and B = B'/B. {Boole also followed this procedure in several of his examples. This gives:

$$\frac{1}{a} = \frac{uB}{b} = \frac{tA}{c} = \frac{tu}{e} = \frac{tuB}{f} = \frac{tuA}{g} = \frac{tuAB}{d+h}$$

Cross-multiplying 1/a by each of the denominators gives six equations:

$$uB = b/a \qquad tuB = f/a$$

$$tA = c/a \qquad tuA = g/a$$

$$tu = e/a \qquad tuAB = (d + h)/a\ .$$

Take logarithms of both sides and put

$$u = \log u; \qquad t = \log t; \qquad A = \log A; \qquad B = \log B$$

and let $\log (b/a) = k_1$, $\log (c/a) = k_2$, . . . , $\log ((d + h)/a) = k_6$ and obtain

$$u + B = k_1 \qquad t + u + B = k_4$$

$$t + A = k_2 \qquad t + u + A = k_5$$

$$t + u = k_3 \qquad t + u + A + B = k_6$$

It is now easy to find two different equations for A and for B. From the 3rd and 5th we get $A = K_5 - K_3$ and from the 4th and the 6th we get $A = K_6 - K_4$. Equating these and replacing the k's by their definitions we obtain immediately that

$$a(d + h) = bc \quad .$$

From the 5th and 6th we get $B = K_6 - K_5$. From the 2nd and 5th we get $u = K_5 - K_2$. From the 1st and this expression for u we get $B = K_1 + K_2 - K_5$. Equating these two expressions for B and replacing the k's by their definitions we find that

$$e(d + h) = fg \, .$$

Substituting the original expressions for these variables gives the two equations that produce Boole's solution to the Challenge Problem.

This procedure makes in relatively easy to find the Boolean solution. It essentially eliminates the algebraic difficulties that led Boole himself to refer to some of his examples as being "repulsive".

APPENDIX D

USING JAYNES' PRINCIPLE

The use we make of Jaynes' Principle is to find the maximum entropy exposure distribution that has the same mean and variance as would be true for the set of magazines we are considering. In addition, the sum of the probabilities must equal one. Generally, we have enough information about a set of magazines to do this. We defined S_1 = sum of the individual magazine reaches and S_2 = sum of the intersections of pairs of magazines. We then observed that S_1 = the mean of the exposure distribution and that $S_1 + 2S_2 = \Sigma i^2 p_i$, the second moment about the origin. What we want to find is the probability distribution with the same mean and the same second moment about the origin and with the maximum entropy of any such distribution.

It is trivially easy to find the desired distribution with Mathematica. I suppose that other programs perhaps may be equally easy to use but I am not aware of any such. Before I knew Mathematica I wrote a program in Basic that would find the maximum entropy distribution but it is simply not worth working out an alternative when Mathematica is today so easily available and is so convenient for a purpose such as this. Therefore, I will describe how to find the maximum entropy distribution using Mathematica.

For the set of magazines under consideration, find S_1 and S_2 and put $L = S_1 + 2S2$. I will assume, as is always true in my experience, that the exposure distribution runs from i = 0 to i = u. Call up Mathematica and write the following definitions on successive lines:

```
p[i_]:=Exp[x+i*y+i^2*z];

F1[i_]:=Sum[p[i],{i,0,u}];

F2[i_]:=Sum[i*p[i],{i,0,u}];

F3[i_]:=Sum[i^2*p[i],{i,0,u}];
```

Then write the single command, where $|S_1|$ means the arithmetical value of S_1 and |L| similarly means the arithmetical value of L:

```
FindRoot[{F1[i]==1,F2[i]==|S1|,F3[i]==|L|},{x,X0},{y,Y0},{z,Z0}];
```

In this command x_0, y_0, and z_0 are initial guesses at the values of the three parameters. Concerning these parameters we know the following: $x < 0$, $y > 0$, and $z < 0$. Any reasonable guess is usually adequate for smaller numbers of magazines, say $n \leq 10$. Thus, for four magazines try $x_0 \sim -1$; for eight magazines try $x_0 \sim -1.5$; for 10 magazines try $x_0 \sim -2$. Look at the values reported for y in the text (Section 37) and it appears that $y_0 \sim .2$ will be generally adequate and $z_0 \sim -.15$.

For the eight-magazine example of Section 37 we have $S_1 = 2.073$ and $S_2 = 2.475$ so $L = 7.023$. Making the initial guesses as described, Mathematica almost instantaneously printed out x → -1.64063, y → .260833, and z → -.162767. Putting these values in the definition of p[i], Mathematica printed out the values given in the exposure distribution in the text.

For larger numbers of magazines it is most important to set Z_0 at a smaller absolute value (but always negative). Thus, I have old data for 49 major magazines. In determining the Jaynes' distribution if all of these magazines were used - for example, during the course of a six-month period - I set the initial values at X_0 = -2.5, Y_0 = .18, and Z_0 = -.004. Almost instantaneously Mathematica gave the values as x = -2.97116, y = .130468, and z = -.0116442, These values produced sums (from i = 0 to 49) of Σp[i] = 1, S_1 = 7.65799, and L = 84.7897. These values produced the following distribution:

i	p[i]
0	.05124
1	.05771
2	.06349
3	.06825
4	.07168
5	.07354
6	.07372
7	.07219
8	.06907
9	.06456
10	.05896
11	.05260
12	.04585
13	.03905
14	.03299
15	.02641
16	.02097
17	.01627
18	.01233
19	.00913
20	.00661
21	.00467
22	.00323
23	.00218
24	.00143
25	.00092
26-30	.00135
31-49	.00009

The smooth regularity of distributions constructed according to Jaynes' Principle together with the solid conservative basis that constitutes the argument for them make this approach a very useful way for the advertising decision maker to compare alternatives in the allocation of the advertising budget.

FOOTNOTES

In these footnotes, LT and SLP are abbreviations for, respectively,

> George Boole: An Investigation of the Laws of Thought, on Which are Founded the Mathematical Theories of Logic and Probabilities; Dover; 1958
>
> George Boole: Studies in Logic and Probability; Watts (London); 1952

CHAPTER I

1. James B. Watson: The Double Helix; Penguin; 1970
2. Michael Riordan: The Hunting of the Quark; Touchstone (New York); 1987
3. The Cambridge and Dublin Mathematical Journal; Nov., 1851 (vol. 6); Reprinted in SLP
4. Problem V_I, Chapter XX; LT; pp. 336f.
5. London, Edinburgh and Dublin Philosophical Magazine and Journal of Science; 1853 (vol. 6); p. 259
6. The Philosophical Magazine; Series 4; 1854 (vol. 7); Reprinted in SLP
7. LT; Problem I, Chapter XX; footnote, p. 321
8. An Investigation of the Laws of Thought; Walton & Maberlay (London); 1854
A Treatise on Differential Equations; Cambridge; 1859
A Treatise on the Calculus of Finite Differences; Cambridge; 1860
9. Reverend Robert Harley: George Boole, F.R.S.; British Quarterly Journal; July, 1866; Reprinted as Appendix A in SLP.
On Boole's life see N. I. Styashkin: History of Mathematical Logic from Leibniz to Peano; M.I.T. Press; 1969; translated from the Russian
10. John Venn: Symbolic Logic; Macmillan (London); 1881; p. 379
11. Charles Sanders Peirce: Collected Papers, Vol. III; Belknap Press, Harvard University; 1933; p. 3
12. Same; Vol. IV; p. 264
13. LT; p.146
14. LT; p. 10
15. LT; p. 15
16. Same
17. Arthur Cayley: "On a Question in the Theory of Probabilities"; Philosophical Magazine; 1853; George Boole: "On the Solution of a Question in the Theory of Probabilities"; Philosophical Magazine; 1854
18. H. Wilbraham: "On the Theory of Chances Developed in Professor Boole's 'Laws of Thought'"; Philosophical Magazine; 1854; R. Dedekind: "Bemerkungen zu einer Aufgabe der Wahrscheinlichkeitsrechnung"; Crelle J.; vol. 1; 1853
19. John Maynard Keynes: A Treatise on Probability; Macmillan (London); 1921
20. Theodore Hailperin: Boole's Logic and Probability; North-Holland (Amsterdam); 2nd ed.; 1986
21. Styashkin, op. cited, pp. 188-190
22. LT; pp. 67-8
23. William W. Bartley, ed. Of Lewis Carroll's Symbolic Logic; Clarkson N. Potter (New York); 1986; See p. 19
24. Bertrand Russell & Alfred North Whitehead: Principia Mathematica; Cambridge University Press; 3 vols.; 1910, 1912, 1913
25. Bartley, op. cited, p. 27
26. Clarence I. Lewis: A Survey of Symbolic Logic, the Classic Algebra of Logic; Dover; 1960; p. 71
27. LT; pp. 279-81

28. Alexander Macfarlane: Principles of the Algebra of Logic; Edinburgh; 1879
29. Charles Sanders Peirce: "On an Improvement in Boole's Calculus of Logic"; in Collected Papers; Vol. III; pp. 3-15; op. cited
LT; pp. 286-7
30. Macfarlane, op. cited, pp. 154-5
Hugh MacColl: "The Calculus of Equivalent Statements"; (6th paper); Proc. London Math. Soc.; 1897 (vol. 28); pp. 113-121
31. J. M. Keynes, op. cited, p. 186
32. LT; p. 338
33. J. M. Keynes, op. cited, p. 431
34. J. M. Keynes, op. cited, p. 432
35. J. M. Keynes, op. cited, p. 433
36. William Feller: An Introduction to Probability Theory and Its Applications; Wiley; 1950
37. William Feller, op. cited, p.236
38. J. M. Keynes, op. cited, footnote, p. 168
39. H. MacColl, op. cited
40. Hugh MacColl: "The Calculus of Equivalent Statements"; (4th paper); Proc. London Math. Soc.; 1880 (vol. 11)
41. Hailperin, op. cited, pp. 366-68
42. J. M. Keynes; op. cited, footnote, p. 167
43. J. M. Keynes; op cited, p. 167
44. J. M. Keynes; op. cited, footnote, p. 186
45. J. M. Keynes; op. cited, much of Chapter VIII; also pp. 164-71
46. J. M. Keynes; op. cited, pp. 167-8
47. J. M. Keynes; op. cited, footnote, p. 187
48. Arthur Cayley: "On a Question in the Theory of Probabilities"; Philosophical Magazine; Series 4; 1862 (vol. 23); pp. 352-65
49. George Boole: "Solution of a Question in Probabilities"; Philosophical Magazine; Series 4; 1854 (vol. 7); pp. 29-32
50. J. M. Keynes; op. cited, pp. 167-8
51. J. M. Keynes; op. cited, p. 188
52. Jorgen Jorgensen: Treatise of Formal Logic; vol. 1; London; 1931; pp 70-201; quoted in N. I. Styazhkin: History of Mathematical Logic From Leibniz to Peano; M.I.T; 1969; p. 189
53. H. Lotze; Logik; Leipzig; 1880; quoted in N. I. Styazhkin, op. cited, p. 189
54. Theodore Hailperin; Boole's Logic and Probability; 2nd ed.; North-Holland; 1986; p. 3
55. Hailperin, op. cited, p. 4
56. Hailperin, op. cited, p. 85
57. Hailperin, op. cited, p. 155
58. Hailperin, op. cited, p. 158
59. Hailperin, op. cited, p. 174
60. John Hooley: "Boole's Method for Solving Boolean Equations"; The Math. Gazette; 1966 (vol. 50); pp. 114-118
61. Hailperin, op. cited, p. 215
62. Hailperin, op. cited, p. 216
63. Hailperin, op. cited, p. 285
64. Hailperin, op. cited, p. 263
65. Hailperin, op. cited, p. 285
66. Hailperin, op. cited, p. 349

67. Note that the last three columns are calculated from the algebraic expressions for b, c, d in terms of "a".
68. G. Polya: "On the Harmonic Mean of Two Numbers"; Amer. Math. Monthly; 1950 (vol. 57); p. 26
69. a = .09 is, in accordance with Polya's argument, the harmonic mean of the minimum and maximum.
70. Hailperin, op. cited, p. 338
71. Cayley, op. cited.
72. Henry Wilbraham: "On the Theory of Chances Developed in Professor Boole's 'Laws of Thought'"; Philosophical Magazine; Series 4; 1854 (vol. 7); pp. 465-76; also in SLP; Appendix B; pp. 473-86
73. Arthur Cayley: "Postscript"; Philosophical Magazine; Series 4; 1862 (vol. 23); p. 471

CHAPTER II

1. Mr. Nassau William Senior (1790-1864) was a well-known English economist.
2. Due to a minor catastrophe in my office, I have lost most of the references to specific logical problems, other than the ones directly used by Boole. Generally, most such problems were gleaned from a considerable number of magazines that carry this kind of challenges to their readers. I apologize to the original authors for my inability to give them credit.
3. There will be further discussion of the relationship between the orthodox theory and Boole's theory of probability in Section 33.
4. George Boole: "On the Application of Probabilities to the Question of the Combination of Testimonies Or Judgments"; Transactions of the Royal Society of Edinburgh; 1857 (vol. 21). Reprinted in SLP.
5. Same; SLP reprint; p. 317
6. Same; SLP reprint; p. 318
7. Same
8. Same
9. J. M. Keynes, op. cited, Chapter V_I
10. J. M. Keynes, op. cited, p. v of preface
11. The chronology is not clear. Boole's Preface to LT is dated Nov., 1853. Wilbraham's article appeared in June, 1854. LT was published in 1854.
12. Wilbraham, op. cited
13. Martin Gardner: Logic Machines and Diagrams; U. of Chicago Press; 1958
14. J. M. Keynes, op. cited, p. 180
15. Hailperin, op. cited, pp. 248-50
16. Hailperin, op. cited, (13) on p. 249
17. Hailperin, op. cited, pp. 250-53
18. Hailperin, op. cited, see Sections 4.7, 5.5, 5.7
19. The equation for "e" can be obtained by starting with fd = bh multiplied out, canceling terms, introducing the expressions for "f" and "g", and further canceling and rearrangement of terms. The equation for "f" starts with f = ed/c written out, substituting fg = eh, and simplifying. The equation for "g" starts with g = ed/b written out, substituting fg = eh, and simplifying.

CHAPTER III

1. Arthur Cayley: "Note on a Question in the Theory of Probabilities"; The London, Edinburgh and Dublin Philosophical Magazine and Journal of Science; Series 4; 1853 (vol. 6); p. 259
2. George Boole: "Solution of a Question in the Theory of Probabilities"; same reference as preceding footnote; 1854 (vol. 7); pp. 29-32

3. Arthur Cayley:"On a Question in the Theory of Probabilities"; Philosophical Magazine; Series 4; May, 1862 (vol. 23, #155); pp. 352f
4. Hailperin, op. cited
5. J. M. Keynes; op. cited, pp. 187-88
6. Henry Wilbraham, op. cited
7. George Boole: "Reply to Some Observations of Mr. Wilbraham on the Theory of Chances"; Philosophical Magazine; Series 4; Aug., 1854 (vol. VIII)
8. ibid.
9. Arthur Cayley; 1862 op. cited
10. George Boole: "On a General Method in the Theory of Probability"; Philosophical Magazine; Series 4; Dec., 1854 (vol. VIII). Reprinted in SLP; pp. 291-307. Quotation on p. 298.
11. Arthur Cayley: "Note"; Philosophical Magazine; Series 4; 1862 (vol. 23); p. 470

CHAPTER IV

1. Henry Wilbraham: "On the Theory of Chances Developed in Professor Boole's 'Laws of Thought'"; Philosophical Magazine; Series 4; 1854 (vol. 7); pp. 465-76; also in SLP; Appendix B; pp. 473-86
2. Arthur Cayley: "Postscript"; Philosophical Magazine; Series 4; 1862 (vol. 23); p. 471
3. George Boole: "On the Application of the Theory of Probabilities to the Question of the Combination of Testimonies or Judgments"; Transactions of the Royal Society of Edinburgh; 1857 (vol. 21). Reprinted in SLP; quotation is pp. 315-16 of SLP
4. George Boole: op. cited, SLP; pp. 308f
5. LT; pp. 251-2
6. LT; p. 262
7. LT; p. 262

CHAPTER V

1. 1960 Consumer Magazine Report; Daniel Starch and Staff, Consultants in Business Research.
2. Sixty-five years under the assumption that the origin of this effort can be dated from M. S. Bartlett: "Contingency Table Interactions"; Journal of the Royal Statistical Society; Supplement 2 (1935) pp. 248-52
3. Stephen E. Fienberg: The Analysis of Cross-Classified Categorical Data; M.I.T. Press; 1977
4. Fienberg, op. cited, p. 32
5. Fienberg, op. cited, p. 57
6. Fienberg, op. cited, p. 59
7. Fienberg, op. cited, p. 73

Printed in Dunstable, United Kingdom

91220872R00121